SPACE

Celebrating 50 Years
Since We Landed on the Moon

Edited by Scott Tilley

Foreword by Winston Scott

SPACE

Front cover photographs from NASA

Back cover photograph from NASA

Published by the Anthology Alliance

Anthology Alliance

An imprint of Precious Publishing, LLC

Precious Publishing

www.PreciousPublishing.biz

ISBN-13: 978-0-9996446-9-0

ISBN 13: 978-0-692-72445-3 (ebook)

TABLE OF CONTENTS

Dedication

For the benefit of all

(NASA)

FOREWORD

Winston Scott

One of the most frequent questions asked of me is, "What is it really like in space?"

The feeling of actually being in space is hard to explain to someone who has never experienced the thrill of lift-off, the freedom of weightlessness, and the awesome sight of Earth from orbit. I trained for thousands of hours before my first shuttle mission, but I was still awed by the whole experience. There's a reason they call space the final frontier: humanity's home looks like a tiny blue marble hanging alone in the vast darkness, surrounded by ... nothing.

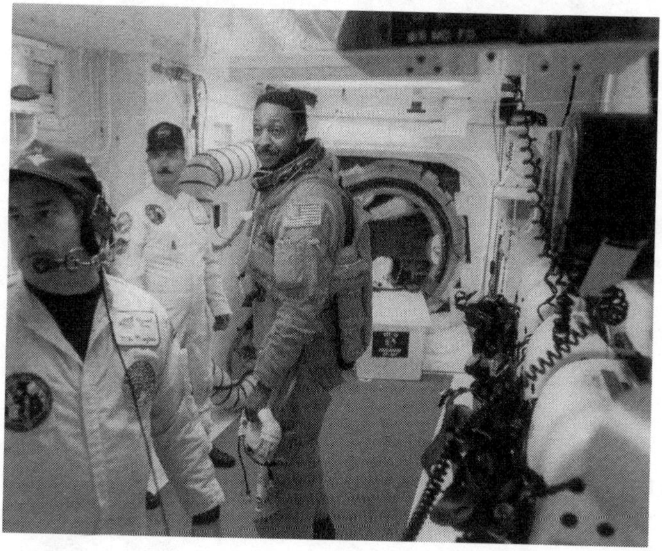

STS-72 (1996): Mission Specialist Winston Scott (center) prepares to enter the Space Shuttle Endeavour at Launch Pad 39B. (NASA)

STS-87 (1997): Like a rising sun lighting up the afternoon sky, Space Shuttle Columbia soars from Launch Pad 39B. (NASA)

I've been interested in space since as long as I can remember. I was fortunate to enter the Naval Aviation Officer Candidate School after graduating from Florida State University in December 1972. This was right after the final crewed mission to the moon, Apollo 17, had flown. Who knows, had I been born a few years earlier, I might have joined Neil Armstrong, Michael Collins, and Buzz Aldrin on Apollo 11. Thankfully, the American space program continued, and I had the opportunity to make my own way into space on Space Shuttle Endeavour in 1996 and again on Columbia in 1997.

As we celebrate the 50th anniversary of landing a man on the moon, it's important not to lose sight of the challenges and opportunities that lie ahead of us. Mars beckons, but to get there (and back) will require many new technological advancements. We've never sent an astronaut so far and for so long, but I'm confident that our collective spirit and imagination can overcome anything that comes our way. The benefits of setting foot on the red planet (and beyond) are hard to fathom, but I've no doubt that the advantages will outweigh the costs, whatever they might be.

Enjoy this collection of stories about space. They cover diverse aspects of our endless interest in the great beyond. The book includes factual accounts of experiences from engineers here on Florida's Space Coast who worked on various NASA missions, fictional musings about our future among the stars, and even satirical tall tales of aliens already here among us.

We've come a long way in 50 years, but we've only just begun our journey into space.

# # #

Winston Scott is a retired United States Navy Captain and former NASA astronaut. He served as a mission specialist on STS-72 in 1996 and STS-87 in 1997. He has logged a total of 24 days, 14 hours, and 34 minutes in space, including three spacewalks. He currently serves as Senior Advisor to the President at the Florida Institute of Technology.

PREFACE

Everyone on Florida's Space Coast has their own story of how the space program has influenced their life. My story began in 1969. I was just five at the time, but I can still remember running home to watch the Apollo astronauts take their first steps on the moon. The TV was an old black and white model, but that didn't detract from the excitement I felt watching live shots from space. It's a testament to how deeply this moment affected me that I can still recall it quite clearly today.

In elementary school and high school, I had several assignments related to space. I often turned to NASA for help, and they obliged every time. I was so excited to receive large envelopes with the blue logo in the mail. I felt as if I, too, was part of the space program. There's no doubt that these projects influenced my career choice to pursue computers and technology.

Fast forward to 1986. Like millions of people, I watched the Challenger disaster live on TV. I can recall this moment in time, too; I was dumbfounded and didn't believe it was really happening. Sadly, it was. I wondered how to build incredibly complex devices like the shuttle so that they could be safer and more reliable. I focus on this topic later in my academic research.

I moved to Florida in early 2003. Shortly after I arrived, I was again glued to the TV, watching aghast as Columbia broke up during re-entry over Texas. Columbia's launch was the first shuttle that I had seen live from the beach. This one felt personal. When Discovery made the "return to flight" in 2005, I watched it live. It was spectacular.

I've felt the rumble of the engines during launch. I've witnessed the flames light up the sky for nighttime launches. I've heard the twin sonic booms as the orbiter hurtles across central Florida. I've seen the shuttle fly piggyback on top of a 747 as it returned home from California. These are all wonderful memories.

On April 17, 2012, I watched Discovery fly into the sunrise over the Atlantic ocean, turn left, and head up the East Coast, bound for Washington, D.C. I thought I was going to feel sad, knowing that this was the very last time I'd ever see the orbiter in the air. In 27 years of active service, Discovery had 39 successful missions. It flew over 149 million miles and spent over a year (366 days) in space. In more ways than one, Discovery led the way. Although I felt a bit melancholy at this bittersweet event, I also felt optimistic about the future of the space program.

It seems my optimism was well-founded. On February 6, 2018, I witnessed SpaceX launch their gigantic Falcon Heavy, all 27 engines roaring. I watched it liftoff via FLORIDA TODAY online, but once the rocket was accelerating away from the pad, I ran outside to see bright orange flames streaking upwards. Not since the final launch of the shuttle in 2011 do I remember feeling so thrilled at witnessing a daytime blastoff.

As amazing as the launch was, the nearly simultaneous landing of the two side cores was even more unbelievable. I could hear the multiple sonic booms coming from Cape Canaveral. Watching the two booster rockets falling from the sky and then settling into perfect landings made the hair on the back of my neck stand up. It looked like an invasion from Mars.

Falcon Heavy reusable side boosters land in unison at Cape Canaveral Landing Zones 1 and 2 following test flight on February 6, 2018. (SpaceX, Wikimedia)

We all know about the many technology spinoffs from the space program that have benefited us. I think it's been a wise investment. But the space program is about more than just business and technology. It represents the culmination of President Kennedy's grand vision. It truly brings out the best in us. It inspired me as a child – and it still does today.

Scott Tilley

Melbourne, FL

September 23, 2019

Acknowledgements

The space program has involved untold thousands of people. They have labored for over 50 years to bring to fruition the dreams and visions of the original program. Without their heroic efforts, space would still be a great unknown. Instead, the next frontier is within our grasp.

Thanks to Cindy Foley for her editing help on one of the submissions. Writers are very creative, and there were some tricky passages to process. Any remaining errors in the text are purely my responsibility.

Lastly, many thanks to everyone who submitted their work to SPACE. Without them, this wonderful anthology would not exist. Ad astra per aspera!

One small step... (NASA)

SPUTNIK: THE START OF THE SPACE RACE

Scott Tilley

We've had an active space program for more than 50 years. NASA was founded 60 years ago, on July 29, 1958. Before the Apollo program, we had the Mercury project. And leading up to Mercury was a series of rocket experiments. In 1926, Robert Goddard launched the first liquid-fueled rocket. During World War II, Wernher von Braun led the development of the V-2 rocket, which in 1944 reached an altitude of 109 miles, making it the first man-made object in space. The V-2 was a genuine threat to London and other allied locations near the end of the war.

But what really kicked off the start of the space race with the Soviets was the launch of the Sputnik satellite by the USSR on October 4, 1957.

Oct. 7, 1957: First Sputnik report (Universal City Studios)

Sputnik was a tiny satellite. It was shaped like a small ball, just over 22 inches in diameter. It weighed barely more than 180 pounds. In *comparison*, the Delta IV Heavy rocket can lift a payload of 63,470 pounds to low earth orbit.

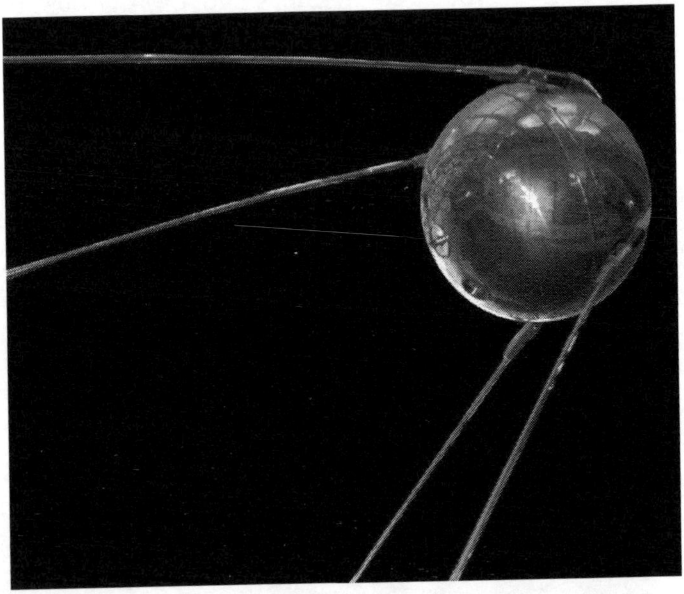

A replica of Sputnik 1, the first artificial satellite in the world to be put into outer space. (NASA)

Sputnik trailed four external radio antennas behind it like tentacles on an octopus. The Soviets did not attempt to hide the satellite's journey through space. In fact, it seems clear they wanted the world to know of their achievement. Sputnik's radio signal was monitored by government agencies and amateur astronomers around the world.

Sputnik's journey only lasted for three months. After three weeks, its batteries died. It burned up on re-entry on January 4, 1958. It had completed 1,440 orbits around the Earth. Its mission was an unqualified success: it announced to the world that the final frontier was now a strategic imperative for the global superpowers. In particular, it heralded the beginning of an unprecedented era of technological innovation, driven by scientists and engineers at NASA and their contractors.

A few years after Sputnik, the USSR launched the first cosmonaut into space. Yuri Gagarin completed one orbit in the Vostok 1 spacecraft on April 12, 1961. The US followed closely behind with Alan Shepard's suborbital flight on May 5, 1961. The space race was well and truly underway.

The launch of Vostok 1 on April 12, 1961 ushered in the era of crewed space exploration. The cosmonaut Yuri Gagarin orbited the Earth once. (NASA)

One significant impact of the USSR's early success in space was a renewed emphasis on STEM education for American children. There was a definite feeling that we were falling behind. How are we doing today?

# # #

Scott Tilley is an emeritus professor at the Florida Institute of Technology, president and founder of the Center for Technology & Society, president and co-founder of Big Data Florida, past president of INCOSE Space Coast, and a Space Coast Writers' Guild Fellow. His recent books include "Systems Analysis & Design" (Cengage, 2020), "Technical Justice" (CTS Press, 2019), and "Dreams" (Anthology Alliance, 2018). He wrote the "Technology Today" column for FLORIDA TODAY from 2010 to 2018. He holds a Ph.D. in computer science from the University of Victoria. Visit his author website at http://www.amazon.com/author/stilley.

Launch of Jupiter-C/Explorer 1 at Cape Canaveral, Florida on January 31, 1958. (NASA/MSFC)

The past is but the beginning of a beginning, and all that is and has been is but the twilight of the dawn.

H.G. Wells, "The Discovery of the Future," 1901

Space for Names, for Who We Are

Christopher Robin Adams

We own a given name for our space and our personhood, and perhaps a nickname, self-chosen or handed to us along our pathway. We may have changed these names or even accepted another name to replace ours or add to it; marriage is the spark for most of these changes. Society provides space for the first and offers more space for the second.

Giving names to people, and things, goes back in time. The standard reference in most of the United States is the Bible: "and G-d created light...and called the light Day..." We feel a kinship with what we name, what we claim to have discovered or created. We birth a child and name her, adopt a child and name him, purchase a car and name it (my parents' first car was Henry, my fourth was Blue). It's what we do. A name is both familiarity and ownership; generally, we don't name what we don't know.

For example, as Jews, we don't name our Creator as it assumes a knowledge which we do not claim to have or possess the capability of having, so typical references might be "G-d" or "Adonai." We capitalize references to The One Who Gave Us Life to emphasize we do not know this One as we know our friends and neighbors, but we do acknowledge Someone was here first, and we respect the mystery and power of that Force.

***

This introduction provides the basis for my names

and their development because a name is both a reference to matter or being itself and to the space that we occupy and to which we are connected. Our name begins our story, and it becomes our place-marker in time, the term for the space we occupy in a continuum. It is the center of the mind map in which we are referenced by memory, voice, and text. We are lost without it ("What is his name? What is his name!") I have five examples resting on this foundation.

Christopher Robin

"Christopher Robin. That sounds nice!" These are the words my mother said when girlfriends were providing possible names for boys as she prepared to deliver me a few months later. Recovering with me in the hospital, she started receiving congratulations cards from near and far. Many referenced "Pooh" or "Piglet" or "Eeyore" or more from the A.A. Milne books "Winnie-The-Pooh", "The House at Pooh Corner", "Now We Are Six", and "When We Were Very Young", all released in the 1920s, well before my birth in 1946.

Surprised at these references, she asked her husband (my father) what was going on. She said he responded with equal surprise, "You didn't know?" So, willy-nilly, I became Christopher Robin. I think that her disconnect from the name triggered the immediate nickname "Kit", which I presume connected to a well-known frontiersman of the day, Kit Carson (really Christopher Houston Carson). He had become historically known for his bravery, later for his suppression of native American tribes, and finally for his support of the Utes. That he could change his mind so completely based on his experiences with the consequences of his actions inspires me to do the same.

Back to Christopher Robin and the space his name occupies in time, and how I represent it in my life. Milne later feared his increasingly popular stories would harm his son and withdrew his association with them; ironically, his son (Christopher Robin, the stories' inspiration) not only became embittered by his name but angry with his father for benefitting financially from the stories at his expense.

Many in the United States met Christopher Robin Milne (CRM) when Sears and Roebuck paid him to go on a tour across the nation, promoting the trademarked brands associated with Milne, Pooh, and so on. According to news reports at the time, CRM was standoffish, aloof, arrogant, rude, and inconsiderate towards the public he met; Sears regretted their promotional decision and reportedly said, "Never again." Ironically, CRM majored in literature and opened a successful bookstore where he regularly dealt with customers who purchased his father's book about him.

At the time Christopher Robin Milne died in 1975, I was thirtyish and beginning to think of my place in time and space. My wife and I had our careers, but I was unsettled. I joined and became active in the Optimist Club where I found an outlet for the community service feelings I had seen operational in my father's whole life. About the same time, I began to believe that I could redeem the name of Christopher Robin, which had reached the United States' shores in such a mixed fashion of books and personality. I told friends that I was going to take actions that would bring back the luster to the name "Christopher Robin." Because of my community service and nice personality, when people thought of "Christopher Robin," they would smile; that idea pleased me, and throughout my life, I believe it has guided me toward the better angels within me.

Kit

I answered to "Kit" from childhood until 1969. It was a short and memorable name, supplanted (at times humorously) by my mother's moniker "Eager Beaver" because of my quick actions to finish work, help around the house, and support family activities. At Wolfie's, my college tuition job in Cocoa Beach, I became The Roadrunner working two shifts in the deli/restaurant to help get my degree in English. The upbeat-ness and positive attitude later fed into the Optimist Club and the goal of redemption for "Christopher Robin."

Christopher

"Christopher," the story and the 20 year space on my timeline, has waited fifty years to be told; now, on the 50th anniversary of the Vietnam War, I will share it.

From the early sixties, in high school, I believed our government's story about Vietnam and our involvement there, though it was barely three years since I (and others) was shocked to learn of our U-2 flights over the Soviet Union. As time passed and our involvement became more questionable in Southeast Asia, my disillusionment deepened. As the draft of young men accelerated to support the war, my frustration began a slow boil, so that on July 1, 1969, as I began serving in the US Army at Fort Jackson, South Carolina, I was a very angry soldier. As my time in the Army became longer, and as I spoke with soldiers returning from the war, my anger increased that I would be asked to take another's life when that person was not threatening me, my family, or my community.

Because of this, I decided with no consultation with family or friends, to divorce "Kit" from the military. I, the personal and personality of "Kit," would no longer serve the military and the process of war; that, as far as it was going to be, would be the job of Christopher. No one in the military knew me as "Kit."

When I was discharged three years later, and began living in Kansas City, Missouri, with my wife, only she and 3-4 very close friends knew of "Kit." All military records, employment records, family records, religious records, coworkers, and personal friends knew me as "Christopher" or "Chris." After my divorce and return to Florida in 1972, I again claimed my childhood name.

When we say "this is my property" or "this belongs to me," we are claiming space. We thus claim a place, an area of geography, or an item of mass whose displacement of air, water, or land indicates a location. It can be pointed to and referred to; language has words for this space and to be able to find this space. We speak of close-talkers as people who step too close to us, who "invade our space" and intrude on the geography around us.

In deciding that I would not share the space of "Kit" that I viewed as my person, my soul, my unique being, I was claiming my full space on earth, the full geography from two feet around me to the center of my being. My parents, family, and friends in Florida knew nothing of this. My two worlds of War and Peace became separated on July 1, 1969.

Why continue as "Christopher" after leaving the military? The war continued. My protests and the thousands of young people who in their own way protested changed nothing; men killing populations became bombs killing populations. I had naïvely believed something would change; nothing did, and my anger continued. "Kit" stayed in Florida; "Christopher" remained working in the Midwest.

Eliezer

Space changed, though, in an unexpected way. I will say, "G-d helped me;" that is the translation of my Jewish name, "Eliezer." While I was enjoying the surprises of new parenthood in 1977 with my wife and son, my father was fighting cancer; he passed in 1979 after a stroke. I had increased responsibility in banking in 1980, but my position was eliminated, and I swore off 3-piece suits and corporate deceit, such that a few months later I was working retail in polo shirts. My father, the parent I found the easiest to talk with, no longer could pick up the phone for my questions.

The space within swirled with conflicting ideas on how to address a world that was changing around me in such drastic ways. The church I attended through high school was no help, the atheism I claimed after that offered no wisdom, and my understanding of Buddhism that nurtured my wife's upbringing appeared beyond my grasp. I reached out to and chose Reform Judaism as a path to the inner peace I sought; the congregants gave me an additional family with which I could connect. Though pain and grief kept intruding on my life and the space around it, I felt I had some balance. I felt that G-d had helped me. It was not the grandfatherly god of childhood, but an amorphous Being Whom I was not compelled to define. Something happened at Sinai; that was enough for me to redefine and rename my space.

Christophe

The last attempt of the outside world to challenge my space has a curious twist. Some technology has named me "Christophe," because the programs those companies use have only 10 slots for a first name, and "Christopher" has 11 letters. Right now, "Christopher" is #57 in popularity for naming new children in the United States, down from #2 in 1994, so no one is changing the computer to meet my name or reach into my space. Therefore, I know that such letters are a grab for money and can be recycled. For humans to connect, at least one must reach out. In the techie world, I am not reaching out to give money to a computer; so as long as "Christophe" is the addressee, my personal space is safe: no time, no pencils, no paper cuts, and no money is wasted. I remain, it seems, untouched from this inhuman corner of society.

In closing, I think this gives me more time for human contact. When I considered teaching for the first time, I was inspired to use the space and time allotted to me to reach others as Sidney Poitier did in "To Sir with Love" in 1967. When I considered teaching for the second time in 1994, I hesitated: would teaching lock me into busy-ness such that I would no longer write? To the contrary, teaching has opened up my life. As students welcomed me into their space, I welcomed them into mine. As they allowed my message to change their work, I allowed their inspired lives to show me that kindness and helpfulness are possible over and over in human life.

After 15 years of teaching, I reflected at the space I turned from, in and after the military. Ten years ago, I wrote about four men in my life who used the military to support their families and strengthen our nation:

Walking Memorial

by Christopher Robin Adams

Walking through the cemetery,
kinda near home,
I brush at the moss hanging from trees,
stare across a canal bracketed with cat tails.
So peaceful.
The soil which cradles those resting here
sustains me,
holds me up,
like my grandfather's words
protecting me in Vietnam's vacuum,
like my father's VMI fight song
delivering me from discouragement,
like my younger brother's kind words
which float down to me like the silken cloth he touched,

like my older brother's wisdom which urged me,
always, to question who we are, as he did, shuttle after shuttle.
Past angers calm a bit;
I kick a sandspur
into the next square foot of Florida.
Whispering Australian pines seem to listen,
crackling oak leaves mark time
back and forth,
back and forth waving
the puppet moss.
I walk and remember.
Grave stones ironically make me laugh
at brighter memories and better times.
The best markers sit within
as I draw courage from these men to step forth,
to speak bravely and write clearly,
to teach passionately and listen thoughtfully,
and to work for a time of intelligent conversation
negating the need for war.

From Christopher Robin to Kit, from Christopher to Eliezer, I have always used my space to work for "a time of intelligent conversation" that would allow us to bridge all chasms between our spaces, to reach out to and embrace others on this journey we travel, and (perhaps, like *The Canterbury Tales*) to seek others who have helped us fill our space with thought and compassion and to thank them for their worthy effort with hopes we might do the same.

May your space be holy.

# # #

Christopher Robin Adams has published writings and shared readings since 1958, both personally and professionally. A retired English teacher (NBCT), he's a National Writing Project mentor, a Space Coast Writers' Guild Fellow, and an officer of Brevard Scribblers. His first collection of poetry, *Spanish Cedar, Preserving the ART of the Cigar Experience,* was released in 2014. His second collection, *Spanish Cedar, Preserving the ART of Courage,* is due for release in 2019.

The second Saturn V launch vehicle for the Apollo 6 mission lifted off from the Kennedy Space Center launch complex on April 4, 1968. (NASA/MSFC)

Man's mind and spirit grow with the space in which they are allowed to operate.

Krafft A. Ehricke, rocket pioneer

THE SEARCH FOR CHALLENGER

Jim Carleton

On January 28, 1986, after being postponed three times and scrubbed once, Challenger (Orbiter Vehicle OV-099) on Mission 51-L, the "teacher in space" flight lifted off the Kennedy Space Center Launch Complex 39B at 11:38 a.m. EST.

I had risen at about 4:30 a.m. and checked the weather. It had been below freezing during the previous night. I dressed for the cold with a three-piece suit and overcoat with lining. In 1986, I was the Chief of Refurbishment Engineering. I worked on the Marshall Booster Assembly Contract (MBAC) that was managed out of Huntsville, Alabama. Our contract, the MBAC Contract, provided the assembly, test, checkout, and refurbishment of all the non-motor parts of the Solid Rocket Booster (SRB). This included the Aft Skirt, Frustum, Forward Skirt, Parachutes, Avionics, Thrust Vector Control (TVC) components, and all other configured parts, totaling 6,584 per flight set. This number does not include approximately 18,445 fasteners. There are two SRBs per flight set. I managed a group of engineers and technicians whose responsibility was to clean, refurbish, repair, repaint, and basically get the hardware back into flight configuration after a launch and retrieval of the two SRBs.

That morning, the temperature at launch was 36° Fahrenheit at ground level. Ice had accumulated at the launch pad area during the evening hours.

The launch of 51-L began at 11:38 a.m. 73 seconds after launch, a massive fireball consumed the vehicle.

The two SRBs flew away in opposite directions and were ultimately terminated by Air Force range safety. The crew compartment with Commander Francis R. Scobee, Pilot Michael J. Smith, Mission Specialist Ellison S. Onizuka, Mission Specialist Judith A. Resnick, Mission Specialist Ronald E. McNair, Payload Specialist Gregory B. Jarvis, and Teacher Christa McAuliffe plummeted into the Atlantic Ocean at 500 miles per hour.

***

"The future is not free; the story of all human progress is one of struggle against all odds. We learned again that this America, which Abraham Lincoln called the last, best hope of man on Earth, was built on heroism and noble sacrifice. It was built by men and women like our seven star voyagers, who answered a call beyond duty, who gave more than was expected or required, and who gave little thought of worldly reward." President Ronald Reagan, January 31, 1986.

On February 3, 1986, President Reagan signed Executive Order 12546, establishing the Presidential Commission on the Space Shuttle Challenger Accident (Rogers Commission).

***

On February 12, 1986, I was requested to support the Harbor Branch Oceanographic Institute from Florida Atlantic University. Harbor Branch was a non-profit ocean research institute operating out of Fort Pierce, Florida. The institute assets included the Johnson Sea Link submersible, capable of reaching more than a half-mile under the ocean and its mother ship the R/V Seward Johnson. The Rogers Commission, through my management, had selected me as the technical representative to dive in the Johnson Sea Link to identify and retrieve Challenger components. As a primary shuttle engineer, I was to categorize Challenger components as opposed to other objects. (Sea Link notes extracted from Jim Carleton's Engineering Log Book.)

While surface craft would identify items on the ocean floor through side scan sonar, our job was to dive to the targets and evaluate whether the object was a Shuttle component or some other object. For some thirty years at that time, rockets, missiles, Mercury, Gemini, and Apollo hardware had been launched off the coast of Kennedy Space Center and the Cape Canaveral Air Force Station. Subsequently, there was a lot of expended equipment laying on the ocean floor.

The Johnson Sea Link, a remarkable submersible, carried a crew of four, in two separate compartments. The aft compartment was originally designed for lockout diving. The forward pilot's compartment was an acrylic sphere with a diameter of five feet, which provided a 360° view in all directions. The sub had Xenon arc lights, still and video cameras, a manipulator arm, and a sonar system.

On February 14, 1986, the R/V Seward Johnson was due at the Port Authority Wharf, southside of the harbor, at Port Canaveral. I was there at 2400 hours. At midnight, I boarded the ship, and at 0045 hours we began the activity to leave Port Canaveral. We had already received coordinates of the proposed impact grid of the right SRB.

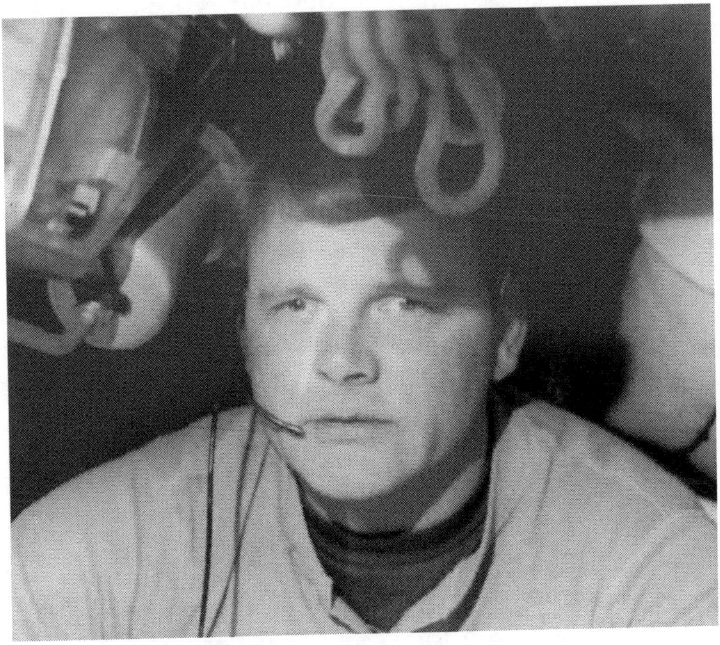

Jim onboard Sea Link, February 1986

On February 15, 1986, at 1330 hours, we started exploring possible targets with the Cable Operated Rescue Device (CORD). This tethered RTV is a video camera controlled onboard the R/V Seward Johnson. We found the first piece of debris which looked like a section of the right-hand Aft Skirt called the kick ring phenolics. Videos were reviewed by all on-board parties, and submersible dives for the Sea Link were scheduled for the next day.

February 16, 1986, 0830 Hours, Dive No. 1244 – Last night, the CORD yielded no more notable video. On board the Sea Link were Tim, John, Phil, and I. We were to proceed to the coordinates provided and take 35 mm photos and video of anything in the sonar indicated area.

0920 Hours – We transitioned into the sea. The R/V Seward Johnson A-frame released us, and we began our first dive. At approximately 900 feet, the initial closed-in feeling had passed, and the scrubbed air felt good. It was totally dark, but at 1,000 feet we turned on the external lights revealing jellyfish and millions of small creatures that were attracted by the light. The bottom of the seafloor was then in sight, covered by sea pens and silt.

0935 Hours – Depth 1,235 feet, visibility 50 feet. We started traveling on course 165 at a speed of 2 knots to coordinates provided. It was tranquil and peaceful, with many crabs and shrimp on the ocean floor. Our speed was reduced to minimize the disturbance of silt.

1010 Hours – We approached a white object. It was an Aft Skirt TVC Hydraulic Frame. We had definitely found a Shuttle debris field.

1015 Hours – I identified a TVC Hydraulic Accumulator, a Fuel Service Module, TVC Frame, exhaust port, nozzle section, and an Aft Skirt Hold-down Post. All these components were from a Space Shuttle Solid Rocket Booster Aft Skirt.

1050 Hours – We next found the completely deformed nozzle section from the Solid Rocket Motor.

1100 Hours – We then employed the manipulator arm to retrieve a component and place it in the attached dive basket attached to the front of the sub. This was the TVC Bootstrap Hydraulic Reservoir.

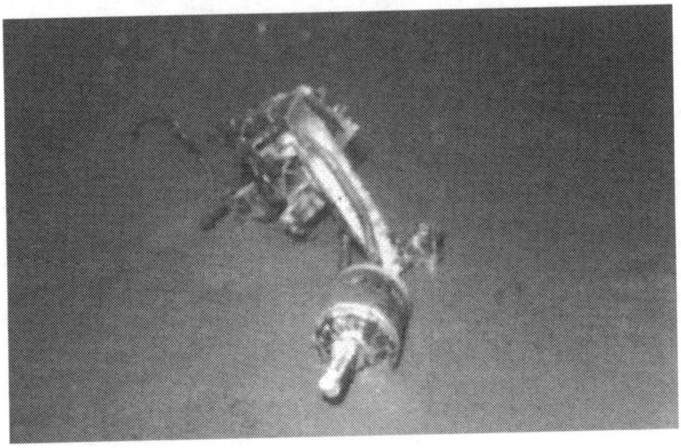

Right hand hydraulic reservoir

1135 Hours – We had seen enough to confirm that this debris field was the Right SRB. The triple motor module was a positive confirmation.

1250 Hours – We continued to photograph and video parts laying everywhere. We had 15 to 20 minutes of battery life left with several targets to explore. Submarine dive time is limited to battery life. We started our ascent. The dive was completed, and we were at the surface to be hoisted back on board the Seward Johnson. The videos taken and the TVC Bootstrap Hydraulic Reservoir with BI026R written on it along with its serial number were positive evidence that we were at the correct location. I had brought configuration documentation on board with recorded information of all SRB components. We then knew we were there. All the photos, videos, and the Reservoir were transferred to a sister ship to return to the port for the rest of the Roger's Commission retrieval personnel to examine.

February 16, 1986, Dive No. 1245 – The second dive of the day yielded no significant information beyond the fact that we were at the correction location.

February 17, 1986, Dive No. 1246 and depth of 1,293 feet – Visibility is 40 feet. Plenty of jellyfish and other marine life. There was a spectacular view, and the area was covered with Aft Skirt and TVC debris hardware. Video recording and photography continued as we mapped and recorded data. We were down for about 4 hours, and I was recording as much information in my engineering log as I could. We started our ascent. I noticed at about 450 feet the light from above was beginning to illuminate the sphere. At 200 feet, we were ascending faster than our bubbles. We surfaced and examined the components we picked up. Some barnacle attachment had already begun.

February 18, 1986, 0830 Hours, Dive No. 1248 – We had transitioned into the sea and were planning to descend to a depth of 1,290 feet. The pilot, Phil, was an experienced diver and sub operator. He was an affable person and a great help in familiarizing me with the Sea Link. I wanted to learn as much as I could about the operation of the vehicle. During this dive, we documented several pieces of nozzle and Linear Shape Charge (LSC). Dive complete. We ascended to charge batteries.

February 21, 1986, 1030 Hours, Dive No. 1250 – We transitioned into the water and were sent on a different mission. This time, instead of looking for Solid Rocket Booster parts, we were in much shallower water (approximately 250 feet). We had been sent to see if we could locate the crew compartment that had been separated from the vehicle during the explosion. We were given codes to report in case we found crew or personal effects. This dive did not reveal anything associated with the crew compartment, so we returned to the ship.

February 22, 1986, Dive No. 1255 – As the dives continued in chronological order, we continued to video and record. But on this dive, we found a wooden-hulled fishing boat. There were hundreds of fish attracted to this artificial reef.

February 22, 1986, 1800 Hours, Dive No. 1256 – Nav Coordinates Latitude 28°36.26', Longitude 80°01.01'.

1950 Hours -- At a depth of 390 feet, we ran into what looked like a DC-3 that we video recorded. We saw the prop of the engine. The engine and wing section stood about 10 feet off the ocean floor, and the sub was overshadowed by the torn-up wing section. The aircraft was in a high state of corrosion. Thirty to forty large fish were using it as a habitat. Some of the fish were up to 350 pounds. Because it wasn't Shuttle related, so we left it and moved to the next target.

It was later determined that the plane was a Grumman TBM Avenger Torpedo bomber. Using the above coordinates, the aircraft was later recovered by a group called Project 19 and the recovery was shown on NBC's "Unsolved Mysteries." Yes, it was an Avenger that may have been from the missing Avenger training Flight

19 from 1945. This dive was over 6 hours, so I was looking forward to food, bathroom and the rest.

February 24, 1986, 1300 Hours, Dive No. 1259 – We were given new coordinates and codes if we discovered the crew compartment. The news media had caused us a lot of additional effort because of the extra security we were forced to employ. The Roger's Commission people were primarily concerned with notifications of relatives and next of kin of the astronauts. We were at a depth of 1,000 feet and beginning our sonar sweeps. The ocean bottom was very clear of shrimp and crabs.

1600 Hours – We located another sunken ship with an outboard motor laying beneath the decaying wooden hull. We could make out the prop and some cables. We left that location and moved forward.

1820 Hours – We located an Aft Booster Motor Segment with kick ring and some Aft Skirt structure attached. I recorded a part number (10182-0092-101) from a systems tunnel. We took many videos of this section of the dive.

1907 Hours – We were finally making our ascent. This was the longest dive so far. I was recording six hours in the sphere, but we retrieved seven or eight good samples of debris. We were looking forward to examining them topside as well as getting some food, drink, bathroom and rest. The R/V Seward Johnson had a fix on us and we prepared to surface and be retrieved by the mother ship.

2000 Hours – The seas were very hostile, and the recovery was arduous. Attaching to the A-frame by the divers took almost an hour.

March 3, 1986, 0821 Hours, Dive No. 1262 – We had permission to dive. This was the first dive of several days due to foul weather and high seas. We had swells up to 13 feet. We landed in the middle of a coral reef. We continued through the coral field with no Shuttle debris noted. We surfaced with no targets acquired.

March 3, 1986, 1830 Hours, Dive No. 1263 – This was my second dive on this day. Several of the crew experienced sea sickness due to the high seas. Once submerged, the top side sea state was not noticeable, and diving was normal. We acquired our first target and headed towards it.

1931 Hours – We located the target, and it was a 55-gallon drum. We proceeded to our second target and found nothing. This ended our second dive of the day.

March 4, 1986, 1447 Hours, Dive No. 1265 – We dived to a new target site, and the depth was about 1,200 feet. This afternoon we placed several Styrofoam coffee cups in the sample basket. I had taken one coffee cup and wrote a note on it to my wife. I wanted to provide a souvenir for her. We were at the bottom at 1,213 feet. The bottom at this depth was covered again with sea pens and various kinds of crabs.

We had seen evidence of one of the other vessels participating in the search and retrieval effort. The Navy Research 1 Vessel (NR-1) atomic submarine was also looking for hardware. The sub has wheels on its bottom and made wheel tracks across the ocean floor as it travels. We had seen these tracks on several other dives.

We did not acquire any new objects this dive.

When we returned on board, we checked our coffee cups. Due to the extreme pressures at 1,200 feet, the cups were compressed to the size between a shot glass and a thimble. The writing was minimized: "To Kathy, 3/4/86 Johnson Sea Link II, The Search for Space Shuttle Debris, Depth 1,200 Ft."

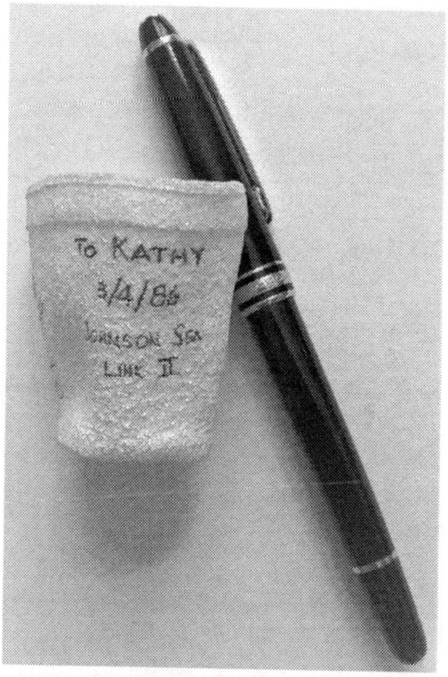

Coffee cup with message to Kathy

March 6, 1986, 0828 Hours, Dive No. 1268 – On this dive, we tracked several sonar targets and found no new Shuttle debris. We did find an outboard motor and several Russian beer cans. Russian trawlers are often seen off the coast during launches. An eight-foot-long hammerhead shark attracted by the light swam by to check us out.

March 7, 1986, 1253 Hours, Dive No. 1272 – This

was my last dive on board the Johnson Sea Link. Most of our task was complete. Based on still shots, video, and samples brought up from the bottom by the Sea Link and all the other resources, the Navy salvagers were able to retrieve the actual piece of Solid Rocket Motor that had the burn-through spot of the motor case, "The Smoking Gun." The rest is history. The Rogers Commission submitted their report on June 6, 1986.

***

After a two and a half year launch shutdown and redesign, STS-26, Shuttle Discovery, the "Return to Flight" mission, was successfully launched on September 29, 1988.

# # #

Jim Carleton retired from the aerospace business after 40 years of service. He supported several missile, rocket, and all 135 Space Shuttle launches. He worked at United Space Alliance (USA), most recently as the Solid Rocket Booster Program Manager. Before USA, Jim worked as a manager for United Technologies Corporation. He retired from the United States Marine

Corps, is a Vietnam Veteran, and a life member of the Disabled American Veterans. He holds a B.Sc. in Mechanical Engineering from the Florida Institute of Technology.

Space Shuttle Challenger atop NASA's Boeing 747, after leaving Edwards AFB in California for the ferry flight to the Kennedy Space Center in Florida for its first launch in 1983. (NASA)

I had the ambition to not only go farther than man had gone before, but to go as far as it was possible to go.

Captain Cook, on his voyage to the Pacific in Endeavor

THE RACE WAS ON

Anne Bonner

"I fear all we have done is to awaken a sleeping giant and fill him with a terrible resolve." *Admiral Yamamotto*, after Japan bombed Pearl Harbor in 1941.

***

USSR launched Sputnik, Oct. 4, 1957 into space.
A small balloon sized satellite that orbited the earth
For three weeks before its batteries died, striking
Palpable fear in the hearts of the free world.
A rallying cry for Americans.
The race was on!

Charles (Charlie) Bassett, II, Captain,
USAF answered the call.
Charlie had been part of the fighter squadron
At Suffolk County AFB, Long Island, NY.
His passion was to become an astronaut
Flying and walking in outer space.

An electrical engineer and a test pilot
Graduating from Aerospace Research Pilot School

And AF Experimental Pilot School, Charlie was
Chosen to be part of the prime crew of Gemini IX.
Elliott See, Cmdr USNR would be the pilot
Charlie, USA's third space walker.

Alas, it was not meant to be.
Cmdr. See piloted the F-38 Talon jet from
Ellington Field near Houston, Texas to
Lambert Field in St. Louis, Missouri.
February's bad weather caused Cmdr. See
To fly into McDonnell's plant's roof, killing
See, age 38 and Bassett, 34, on Feb. 28, 1966.

A rallying cry for military wives.
The first to respond to the tragedy
To aid and support Charlie's wife, Jeannie
And their two children.

Could it have been worse?
Losing two brilliant loving men was grievous
However, hundreds of workers could have been
Killed, the Gemini spacecraft ruined and
America's space race set back for years.

\# # #

Anne Bonner, a 5th generation Floridian, born and raised on the Space Coast, has written ten historical fiction books: Five young adult books and five love stories. She earned the NSDAR Award, "In Recognition of Outstanding Achievement in Historic Preservation" for her historical writings and workshops in schools. Anne is a past member of the Boards of the Brevard Heritage Council and the Space Coast Writers' Guild, and is a member of American Pen Women.

A nearly full moon sets as the Space Shuttle Discovery sits atop launch pad 39A at the Kennedy Space Center in Cape Canaveral, Florida on Wednesday, March 11, 2009. (NASA/Bill Ingalls)

Flight by machines heavier than air is unpractical and insignificant, if not utterly impossible.

Simon Newcomb, 1902 (18 months before Kitty Hawk)

SPACE: A LAUNCHPAD FOR AFRICAN AMERICAN ACHIEVEMENTS

Art Edwards

The Final Frontier

"Space – The Final Frontier" is the beginning of a monologue from one of the most popular television shows of the 1960s. This famous phrase from the original "Star Trek" is universally recognized and conjures up images of futuristic space travel; it has stoked the imagination of generations. Although the original series only aired for three years, "Star Trek" gained a cult following and survived for decades in syndication. In 1987, the franchise was revived with new characters in "Star Trek: The Next Generation." This successful series was followed by "Star Trek: Deep Space Nine" in 1993 and "Star Trek: Voyager" in 1995. The next millennium saw the introduction of "Star Trek: Enterprise" in 2001 and "Star Trek: Discovery" in 2017.

The National Aeronautics and Space Administration (NASA) began operations in 1958 with about 8,000 employees and a $100 million annual budget. NASA operations were headquartered in Washington, DC. NASA operational sites consisted of three research laboratories, including Langley Aeronautical Laboratory in Virginia, Ames Aeronautical Laboratory near San Francisco, and the Lewis Flight Propulsion Laboratory near Cleveland. In addition to the three research laboratories, NASA operated a high-speed flight research facility in the California desert and a facility for developing and testing rockets at Wallops Island, Virginia.

In a speech to Congress on May 25, 1961, President John F. Kennedy unveiled the vision of the Apollo Project and the goal of landing a man on the moon: *"I believe that this nation should commit itself to achieving the goal, before this decade is out, of landing a man on the moon and returning him safely to the Earth. No single space project in this period will be more impressive to mankind, or more important for the long-range exploration of space; and none will be so difficult or expensive to accomplish."* Thus, America's commitment to the Space Race had begun. Over the years, NASA has led various space exploration programs, such as the Apollo moon landings, the development of the space shuttle, the incredible Hubble telescope, and the unprecedented scientific orbiting scientific laboratory – the International Space Station (ISS). These and numerous other NASA programs have been critical in advancing technology and improving our everyday lives, moving us ever closer to the idyllic world depicted in "Star Trek."

We Shall Overcome

The 1960s also gave rise to another well-known phrase: *"We Shall Overcome."* For many years, African Americans struggling for equality and civil rights found solace and strength in this phrase. Ironically, "We Shall Overcome" was a key phrase in a historic speech to a joint session of Congress in 1965 by President Lyndon B. Johnson. This speech is especially memorable because in pleading to Congress to pass voting rights legislation, President Johnson became emotional and openly wept during his speech. President Johnson not only convinced Congress to pass the historic Civil Rights legislation, but his unrelenting efforts influenced and drove a commitment to inclusion of Civil Rights as a priority in NASA space programs.

In addition to well-known scientific and technological advancements, the space program has been a significant factor in lifting numerous African Americans out of poverty. Moreover, the space program has provided a launchpad in the successful careers of numerous African Americans and the ascension of African Americans in the business world.

Through conscious and deliberate efforts, NASA has been instrumental in assisting bringing about a measure of social and economic change. NASA directives mandated equal opportunity in hiring and promotions at NASA facilities. Many of the NASA contractors complied with the NASA mandates in their federal contract work, while also adopting an affirmative action approach at non-NASA worksites.

NASA, through federal mandates, drove the hiring of African American men and women throughout the United States. The significance of NASA's equal opportunity mandates played a key role in the creation and expansion of opportunities for African Americans living in southern states where many large NASA facilities are located. For example, in locations such as the John F. Kennedy Space Center in Florida, the Johnson Space Center Houston in Texas, the Marshall Space Flight Center in Alabama, and the John C. Stennis

Space Center in Mississippi all employed thousands of workers (including many African Americans).

The workforce at these and other NASA Facilities included many workers who migrated from other parts of the country, as well as many workers from the local areas. The employment at the facilities allowed African Americans previously employed in low-paying agriculture and other domestic jobs to work in less physically demanding environments while earning higher wages. More importantly, work at these locations allowed many African Americans to showcase their abilities and talents on an unprecedented level.

My Story

Without a doubt, many African American pioneers working in America's space program helped break down cultural and social barriers and to change the perception of black people in America – specifically in the south. This is true for me as well. I was born in a very small town (Marianna, Florida), where I was raised on my grandparent's 160-acre farm. On the farm, I learned many "soft skills," such as how to manage money, how to set goals, and the importance of being dependable and on-time. These (and many other) attributes that are not taught in college but are invaluable in both the workplace and at home.

My secondary education was at Jackson County Training School (JCTS), where I entered 1st grade in 1954 and completed 12th grade in 1966. The school was in Marianna and was 10-15 miles away from our farm. I rode the bus to and from school every day. My cousins and I had to walk several miles to the bus stop, and we never missed the bus. In the segregated Marianna school system, blacks attended JCTS and whites attended Marianna High School (MHS). I cannot remember ever having a new book at JCTS; every book that I received had been used by other students. Usually, we had hand-me-down books from MHS.

In my teenage years, I developed a passion for electronics. I can recall spending most of the money that I earned (and some of my lunch money) on transistor radios. I purchased quite a few and eventually tore them apart to learn how they worked. Sometimes I successfully put the radios back together, and sometimes they didn't work. But each transistor radio provided a learning experience.

I laugh now when I remember using a fire poker as a soldering iron. What a mess! Even the crude soldering iron that I finally crafted didn't do the trick; I still had to use the fireplace or our wood burning stove to heat up my soldering iron. My love for electronics was vital in my enrollment at Florida A&M University (FAMU) following high school graduation

Electronics and career plans led to eventually have my own business. The second month of my college classes, the Florida A&M University (FAMU) Dean of Engineering, Col. Blake, persuaded me to change my major to electrical engineering. As a result of a civil service exam test score, I was selected to participate in a NASA Engineering Co-Operative Education program at Kennedy Space Center (KSC). In April 1967, I reported to work at the KSC Central Instrumentation Facility

(CIF) Antenna Site. The CIF was a rocket tracking station with hands-on work. We designed and built test equipment, maintained tracking equipment (including the large dish antennas), and tracked missiles through the first stage of flight.

I worked as an engineering co-op student from 1967-1969, earning enough money to pay for most of my college education. The CIF Antenna Site offered a tremendous learning experience. When I graduated from FAMU in December 1970, I had quite a few job offers. I can vividly remember the first airplane ride (of my life) to a job interview with AT&T in the Wachovia building in Greensboro, North Carolina. The AT&T operations in Greensboro was in the area called "The Triad," consisting of Greensboro, Winston-Salem, and High Point. The Triad was a very impressive area, and in the late 1960s it was becoming a business and technology hub. AT&T made me a very tempting job offer, but I had developed a passion for the excitement of the space program.

My Aerospace Career

I ultimately spent over 45 years working in the nation's space programs, primarily supporting human-crewed flight at KSC. My very diverse career included electrical/avionics engineering, software engineering, and safety. My career progressions included management assignments that culminated in an executive management position as the director of the space shuttle Launch Processing System (LPS).

A rainbow breaks through the clouds behind the Vehicle Assembly Building at NASA's Kennedy Space Center (KSC) in Florida. (NASA/Jim Grossmann)

As a shuttle avionics lead engineer, one of my responsibilities was electrical/avionics design support for the space shuttle Remote Manipulator System (RMS), which was scheduled to fly on the second space shuttle flight. The 50-foot RMS, commonly called the shuttle robot arm, was a pivotal addition to the shuttle. The arm was necessary for deploying and retrieving satellites from the space shuttle cargo bay, and for constructing the ISS.

After the Challenger tragedy, I accepted a promotion from supervisor of the shuttle Data Processing System (DPS) to the manager of the LPS Applications Software Department. After the Challenger disaster, all the shuttle checkout and launch software was rewritten and re-platformed – on time and under budget. I was subsequently promoted to LPS director, where for over five years I provided leadership in preparing the ground processing computer systems for shuttle test and launch support.

Secured on a foot restraint device connected to the Canadian-built Remote Manipulator System (RSM) aboard the Earth orbiting Atlantis, astronaut Jerry Ross participates in a STS-61B demonstration of future space station construction. (NASA)

***

I consider improving KSC employee safety through the OSHA Voluntary Protection Program (VPP) to be one of my most significant achievements. I provided vision and leadership in effecting a culture change to a United Space Alliance (USA) employee workforce of over 8,000 people, housed in over 500 facilities at the Space Center. In 2001, USA at KSC became the largest OSHA VPP worksite and was honored with Star Certification.

Very few worksites qualify for this prestigious program. VPP Star Certification resulted in measurable reductions in safety-related incidents, employee injuries, and lost-time cases. The safety cultural changes extended beyond shuttle processing, as workers took their safety consciousness home and increased safety awareness and safety practices throughout the community.

Indeed, America's space program provided a path for me, a poor but ambitious small-town farm boy, to excel and prosper. KSC was my career launchpad.

Kennedy Space Center

Another popular television program from the 1950s closed with the famous line, "There are eight million stories in the naked city; this has been one of them." Likewise, my story is just one of thousands illustrating how African Americans benefited from the space program as a launching pad for civil rights and their careers. Furthermore, my story is localized to African Americans working at Americas' Spaceport: Kennedy Space Center.

KSC remains NASA's primary launch center of human spaceflight. It is located on what was once the northern portion of Merritt Island, Florida. It is bounded by the Indian River to the west, the Banana River to the south, and the shores of the Atlantic Ocean to the east. Until the early 1950s, Merritt Island was a residential area where people earned their living primarily through agriculture (farming and citrus groves) and fishing. In the mid-1950s, NASA acquired acres of privately-owned land for construction of KSC facilities. Altogether, KSC covers over 144,000 acres.

My wife, Audrey, who grew up in Titusville, Florida, has shared with me countless stories about the impact of KSC on the local community and, more importantly, on the local African American community. She recounts the pre-KSC economic struggles of the North Brevard community in general and the African American Community in particular. The citrus and agriculture businesses in Brevard County attracted workers from throughout the southeast (e.g., Alabama, Georgia, Mississippi, South Carolina). Many seasonal workers from the Caribbean who were proficient at harvesting oranges were also attracted to the area. In fact, in the north Brevard area of Mims, several labor camps were built specifically to house the migrant workers.

Audrey recalls her favorite aunt, Carrie Brown, who worked as a domestic, found herself unexpectedly unemployed. This was not because of her work or work attendance; Aunt Carrie was laid off at the convenience of her employer. What seemed like bad luck eventually worked in Aunt Carrie's favor. She was hired on the KSC janitorial staff and spent over 30 years working there. She was an excellent worker and highly regarded for her logic and common sense. She was a labor activist and promoted fair treatment and fair pay for janitorial workers.

***

The movie "Hidden Figures," released in 2017, tells the story of a group of female African American mathematicians working at Langley Air Force Base in Virginia during the early years of crewed space flight. These ladies, referred to as "human computers," were adept in using manual devices, such as slide rules and adding machines, to perform complex calculations needed for space craft Guidance, Navigation & Control (GN&C). These ladies famously contributed to brainstorming and solving perplexing GN&C problems.

Coincidentally, around the same time, several young African American mathematicians were performing complex computations at KSC. Young scientists from Historically Black Colleges & Universities (HBCU's), such as Lonnie Blocker (a graduate of Bethune Cookman College), Herb Pete (a graduate of North Carolina A&T), and many others were making their mark at the space center.

Success Stories

African Americans have made many notable contributions at KSC. At the height of the space shuttle program, KSC employed a workforce exceeding 20,000 across hundreds of disciplines. There are thousands of African American "space" success stories spanning the past 50 years. Three of these stories concern James (Jim) Wilder, Napoleon (Nap) Carroll, and Hugh Brown.

James (Jim) Wilder: Technical & Leadership Excellence

In 1968, James (Jim) Wilder arrived at KSC to work

as an engineering co-op student. He was hired by Rockwell Aviation, the company responsible for the Apollo Command and Service Module (CSM). During space missions, the CSM provided a habitat for Apollo astronauts and provided storage space for equipment and items such as moon rocks. One of the most critical CSM systems was the propulsion system needed for atmospheric re-entry.

This view of the Apollo 9 Command and Service Module (CSM) was photographed from lunar module "Spider" on fifth day of the mission. (NASA)

After a few co-op periods, Jim gained enormous respect from everyone who he encountered: engineers, technicians, management, and NASA customers. He quickly grasped the intricacies of the Apollo CSM hydraulics system, which was a critical component of the CSM propulsion system. After graduation from Florida A&M University in 1971, Jim was hired as a CSM hydraulic systems engineer. Over the next 40 years, Jim continuously demonstrated superior technical excellence, coupled with an outstanding ability to plan and execute technical projects and to lead others.

On the shuttle program, the KSC engineering team had hands-on participation in the development of shuttle ground checkout software. This participation ranged from creating requirements and specifications to more complex activities, such as creating executable computer code. Many of the systems engineers on the shuttle team relied on programmers to develop the actual computer code. Complex shuttle systems required engineers to understand the systems completely, and to also understand the capabilities and limitations of the Launch Processing System (LPS). As the lead engineer for the shuttle hydraulics system, Jim had the responsibility to ensure the creation of needed test procedures and also the development of LPS test checkout software. I was impressed by Jim's ability to grasp the computer programming needed to test the system.

Throughout the shuttle program, Jim Wilder demonstrated superior technical and leadership performance. His abilities were recognized and appreciated by management and the NASA customer as he ascended from lead shuttle hydraulics engineer to fluids management positions, shuttle project office director, director of the NASA logistics depot (NSLD), and shuttle ground operations deputy associate program manager. His last role was United Space Alliance VP of Orbiter Systems, where he oversaw all aspects of shuttle orbiter preparations. Most importantly, he made the "Final Go-For Launch" call for the space shuttle orbiter.

Napoleon (Nap) Carrol: Financial & Leadership Excellence

Napoleon (Nap) Carroll started his career at KSC in 1972 as a co-op student from Bethune-Cookman College. During his co-op periods, he worked in the KSC Resources Management Office, the office responsible for developing and managing KSC's budget. After graduating with a degree in Business Administration, Nap was hired as a budget analyst in the KSC comptroller's office. He spent time supporting the office, managing the center's resources for facilities development and management, multiple program management funding, design engineering, vehicle test and check out, and base KSC operations.

Nap's job performance, coupled with his tactical thinking and excellent people skills, was noticed by management as well as his peers. His rise in NASA began with his assignment as chief of the Shuttle Engineering Directorate Resource Management Office (SDRMO), supporting the KSC shuttle directorate, as well as the NASA shuttle logistics depot located off-site in Cape Canaveral. This assignment carried great responsibilities in managing and administrating shuttle program resources and providing accountability to the KSC center director and NASA headquarters. In 1996 and 1997, Nap served as the acting Chief Financial Officer (CFO) at Stennis Space Center in Mississippi. In 1998, he was promoted to CFO at KSC, where he and his staff had responsibility for all aspects of the KSC budget.

Although his strengths were in the world of business and finance, Nap was a tactical thinker with the ability to grasp technical details. I especially remember his approach to "bid whist" games at work. He was a keen observer and deliberate thinker in these competitive lunchtime breaks, where many KSC African Americans gathered to socialize, compete, and offer mutual support. In my mind, Naps ability for tactical thinking played a key role in his success as the executive director of the Cape Canaveral Spaceport Management Office (CCSMO) in administration of the joint base operations contract, which provided services at both KSC and the adjacent Cape Canaveral Air Force Station. In the CFO position, and in the position as CCSMO executive director, Nap reported directly to the KSC director.

Throughout his career, Nap always offered a helping hand to others seeking advancement. His success led to opportunities for many of his co-workers.

Hugh Brown: Entrepreneurial Success

In the infancy of the space program, before the advent of supercomputers, a workforce with specialized analytical skills was needed to retrieve intelligence from telemetric data from ground processing systems and flight systems. In 1968, Hugh M. Brown was hired by ITT Federal Electric Corporation as a telemetric engineer supporting the Apollo program. Hugh's degree in Mathematics and Mechanical Engineering from Central State University in Ohio uniquely qualified him for the daunting task of analyzing downlinked Apollo systems data. This work assignment provided the opportunity for Hugh to distinguish himself as a valuable asset in the complex world of data synthesis.

In 1978, Hugh yielded to his entrepreneurial spirit and founded his own engineering and technical services company, BAMSI. The unique thing is that he started what would become a very successful company in his garage. Under Hugh's leadership and with his vision, BAMSI began as a NASA KSC contractor and became one of the most successful privately-owned firms in the southeast. Federal documents reveal that the company employed more than 2,200 highly skilled professional and technical personnel in 11 states, achieving annual sales of more than $90 million in just over 18 years.

Hugh's capabilities have been recognized at the local, regional, and national levels. He served on the corporate boards of Blue Cross & Blue Shield, and SunTrust Bank of Orlando. He has also served on the Governor's Commission on Education and as Chairman of the Florida Space Authority. In the late 1990s, Hugh received an appointment to the Southeast Federal Reserve Bank, which in and of itself was a notable achievement. His subsequent rise to the chairmanship of the Federal Reserve Bank of Atlanta and Federal Reserve Bank of Jacksonville is historic.

Hugh has been the recipient of numerous awards and recognitions. For example, he is a four-time recipient of the KSC Small Minority-Owned Business of the Year. In 1991, at a White House ceremony, President George Bush presented Mr. Brown with the U.S. Small Business Administration's 8(a) Graduate of the Year award.

In the early 1970s, I vividly remember serving with Hugh Brown on the Board of Directors of the Brevard County Community Services Council. I greatly admired his poise and his common-sense approach to working issues and solving problems. Over the years, Hugh has made community and public service a priority; many people have benefited from his philanthropy.

Hugh has been (and continues to be) a mentor and an inspiration for many African Americans. Indeed, he has been an inspiration and a mentor to many throughout the United States. I recall the day in the late 1980s when Ken Pinson, the first African American Sears store manager in Brevard County, told me that he was leaving Sears to start his own business. I greatly admired Ken and had a concern about his move, but I was heartened to learn that Hugh would be advising Ken as he made his foray into entrepreneurship. In 1989, Ken Pinson established Genesis VII, which became a successful procurement company serving NASA and other companies in Central Florida. I recognize that the success of Genesis VII was due in large part to the efforts of Ken Pinson, but the value of having a successful businessman such as Hugh Brown as a mentor cannot be overstated.

Summary

"Space - The Final Frontier" is an apt description of

the human exploration beyond Earth's surface. The American space programs have provided the opportunity for the country to capitalize on previously ignored talent. Just as "there are millions of stories in the naked city," there are thousands of stories (told and untold, known and unknown) that Illustrate how African Americans have benefited from space.

# # #

Art Edwards is President and CEO of Occupational Safety Management Safety Specialists, Inc. and Edwards Media Group, Inc. He supported crewed space flight at the Kennedy Space Center for over 45 years. His career included various engineering and management assignments. His awards include a National Space Club "Life Time Achievement Award." Art has a B.Sc. in Physics and an M.Sc. in Industrial Engineering. He is a Certified Occupational Safety Specialist and a Florida State Certified General Contractor.

May 25, 1961: President John F. Kennedy, with Vice-President Lyndon B. Johnson and Speaker of the House Sam Rayburn behind him, addressing the Joint House of Congress. (NASA/JSC)

It is difficult to say what is impossible, for the dream of yesterday is the hope of today and reality of tomorrow.

Robert Goddard

Rendezvous Among the Stars

Joanne Fisher

Nadia was strapped in her seat and preparing for lift-off. No matter how much training she had endured, butterflies filled her stomach. This was the real thing. She had been raised to control her emotions, guard them against others, and to be practical. The silence inside her helmet was deafening. "Why am I feeling this way?" she asked herself. But her mike was hot, and her colleague heard something.

"Did you say something, Nadia?" asked Chandler, who was the only other person on board.

"No, no, I'm fine. Just a bit nervous."

"Trust me; we're all in the same boat."

"Don't you mean the same spacecraft?" she asked.

"Touché!"

Nadia and Chandler had trained long and hard for this very day. She had gotten to know him well, and she liked what she saw. He was a typical American: rather tall, with light brown hair, a slim, muscular build, and piercing blue eyes. Plus, he had a charming personality, unlike what she was taught to believe about American men. Under other circumstances, she would have enjoyed a fling with him, but that was forbidden by policy – and especially by Mother Russia.

"Ten. Nine. Eight. Seven. Six. Five. Four. Three. Two. One! Ignition and lift-off! We have lift-off of the Eagle

Space Vehicle on its first crewed trip to the International Space Station!"

No training was able to adequately prepare her for this. Her nervousness increased and jolted her out of her comfort zone as the roar of the engines became louder and more physical, rattling her to the core. Nadia closed her eyes and held on tight to the seat handles. As the rocket carrying its human cargo began moving upwards into the clouds, she said a silent prayer in the hopes that it wouldn't have a "catastrophic failure," as the Space Agency was wont to say. Somehow, it seemed a bit tamer than the word "explosion." The "catastrophic failure" of 2020 was a very sobering event. Fortunately, that spacecraft was carrying only electronic equipment, food, and supplies, but it was still a devastating moment for Russia. The Space Matryoshka Team had to work extremely hard to repair the "anomalies" (another Space Agency term) and bring their launch vehicle back to perfect working condition.

"Oh, Lord, please let this vehicle arrive at the ISS safe and sound," she whispered to herself.

"How are you doing, Nadia?" Chandler had to speak loudly and distinctly to be heard over the din of the rocket's engines. This was the most powerful rocket that either the United States or Russia had ever launched.

His low-pitched voice was comforting. "So far, so good," she yelled back.

About four minutes into the flight, the boosters disconnected, the main engines shut down, and the spacecraft silently glided into space. Nadia marveled at how the vibration and noise abruptly stopped, and the roaring silence again filled her helmet. The two passengers were suddenly weightless. Now came the game of chase as their spacecraft hurtled toward the ISS. With no frame of reference, they didn't perceive the speed at which they were traveling. Their senses told them that they were motionless. In fact, they were going more than 17,000 miles per hour.

They unbuckled their belts and began to float like balloons do when released into the sky. Since they had several hours until the rendezvous and docking with the Space Station, they slipped out of their pressurized space suits and floated about the small spaceship. This was much easier than in the early days of both countries' space programs, back in the late 1950s and into the 1980s – long before either of these two spacefarers were born, when the United States began its Space Shuttle program. In those days, it had taken several men to assist the astronauts and cosmonauts in donning their suits. Because the Shuttle and the ISS stayed in orbit for extended periods of time, lighter and thinner suits were developed that allowed astronauts to do the job themselves.

"You know, Nadia, even though I've been thoroughly trained and been on several missions, I'll never get used to this."

"Oh yes, same here."

"How long have you trained?" he asked. "Before we began training for this mission." He thought she was drop-dead gorgeous. She was what you might expect a

Russian beauty to be: long blonde hair, emerald eyes, slim build, but with a beautiful bosom and those long legs! She was irresistible!

"Two years."

"I see. And if I may ask, what were you doing before you became an astronaut?" He was prepared to hear, "I was a model."

"I was teaching physics at the University of Moscow."

He absolutely adored her accent. "Wow! You're a university professor! I'm impressed!"

"Thank you, Chandler." She loved to receive compliments, especially from this handsome American. She stretched her legs and arms and then made herself comfortable in her seat. She latched the belt and pulled out a book to read. They still had over five hours before they could dock with the Space Station, and reading was her favorite way of wasting time.

Chandler was impressed by this beauty. Beautiful and smart! The perfect woman! He, too, stretched his limbs and relaxed in his seat. Instead of reading, he flipped the seat back and shut his eyes.

Ah! He's a sleeper, she thought. *I don't like lazy people!* She shrugged her shoulders. *Maybe that's the way Americans are. I shouldn't judge.* A few hours into her book, she too, fell asleep and her book slid off her lap and began floating into the air.

Something hit his helmet ever so lightly, and he woke up to see a book floating next to his face. He looked over to the beautiful blonde sitting in the next seat and smiled. *She looks like an angel.* He grabbed the book and placed it in the pocket on the side of her seat and pressed the Velcro to seal it.

"Okay, sleepy heads, we're about to rendezvous. Begin your docking preparations."

They both awoke and began running down the checklist. Each of them had their own tasks, and they completed them smoothly and accurately, just as in training.

The docking to the ISS sounded like a tiger who had awakened and was roaring with hunger. Even with all the training they both had, that sound gave them a start.

One hour later, they were meeting with the ISS Mission Commander. He had been at his post for the past two weeks and they would be under his command for the duration of their mission.

"Good afternoon and welcome to the International Space Station! I am William Cedrick Lollows, but you can call me Buddy." He reached over to shake Nadia's hand.

Buddy was a decorated colonel, a "full bird" in unofficial military parlance, describing the eagle insignia on the collar of the uniform. He had spent several years in Special Forces. Buddy had always been fascinated with the shuttle program, and so he decided to join the American Space Force at the dawn of this new branch of the military.

"Very nice to make your acquaintance, Buddy." She

shook his hand and gave him a warm smile.

"You must be Nadia Katarina Subciek, or better yet, Professor Subciek?"

"Yes, you are correct."

"And you must be Chandler Robert Barrett, mechanical engineer, is that right?" he shifted his hand to shake Chandler's.

"Yes, sir, you are correct. Nice to meet you, sir."

"Oh, I'm no sir! Please call me Buddy."

"Okay, Buddy. So, where are you from in the States?"

"Oh, you hear my accent. Buford, Georgia. You?"

"Hartford, Connecticut."

"New England, huh? I'll tell you, son, those Patriots never cease to amaze me!"

"Yes, they're really something!"

"I still don't know how they pulled off that win over the Falcons!"

"Oh, yes! That was some game!"

Nadia was beginning to become bored at that conversation, and Chandler noticed her yawning from the corner of his eye.

"Okay, Buddy, can you give us the 4-1-1?"

"Oh, yes, of course! Please follow me into the conference area."

Buddy began floating into a chamber of the station that was located to the left and then the right of it. As he floated along, he grabbed handles that were found throughout the space station to guide his movement. Chandler and Nadia followed him in the same manner.

It doesn't take long to adapt to living in space. It took these new astronauts a day or so. Zero gravity makes life drastically different from Earth. The first thing a visitor to the ISS would notice is that there is no up or down, no floors or walkways. You float to wherever it is you're going. Speaking of going, that's all different, too. You have to strap yourself to the toilet to create a tight seal. Then a vacuum pump draws everything away from you. This is placed into a cargo rocket that is jettisoned, or released, to fall back into the Earth's atmosphere and burn up. The lack of floors or ceilings makes organization quite easy. All four sides are composed of storage units, or work tables, for experiments. One other benefit of Zero G is that if you're writing in a notebook, you don't have to put down your pen. Just let go of it, and it'll continue floating right there where you left it!

Back in the "old days," food was reduced to a thick paste and packaged in something resembling a large toothpaste tube. The astronaut would simply squeeze the tube of goop into his mouth. It was done this way because in the early days they didn't want food particles or liquids floating into the electronics panels and possibly resulting in a catastrophic failure. As time went on, and spacecraft design improved, food began to be vacuum-packed. Astronauts became adept at catching food crumbs. This allowed food to taste much better. Dirty clothing is placed into a container and returned to

Earth, along with other items intended not to be disposed of. Each astronaut has his or her own sleeping quarters. It consists of an elaborate sleeping bag hanging on the wall. The astronauts zip themselves into their bags to prevent them from just floating around.

One afternoon, Chandler was at the amateur radio setup and was having a conversation with what seemed like children. Nadia was intrigued by this activity and decided to join him.

"Well, Scott, here comes Nadia, our beautiful Russian professor. Would you like to say hi?"

An excited voice crackled in the speaker: "Yeah, sure."

Chandler handed the microphone to Nadia. "Just push and talk," he said with a coy smile.

"Hello, this is Nadia. Who's this?"

"This is Liam. Hello Nadia. Are you having fun up there?"

Nadia chuckled. "Oh, yes! Lots of fun." She looked at Chandler and shrugged.

"Oh, Miss Nadia, I saw you on TV. You're really beautiful!" A different voice was heard.

"Well, thank you very much. And what is your name?"

"I'm Vanessa."

"Hello, Vanessa. What grade are you in?" She was beginning to enjoy this activity.

"Fifth. And you?"

"I'm not a student. I'm a professor."

"Oh, wow!"

"Miss Nadia, my name is Kyle, I'm in fifth grade, and I want to marry you. Please?"

Chandler was enjoying every minute of this and especially now that Nadia had received a proposal. He gave her a look like, "You need to give him an answer."

Nadia had a horrified look on her face. "What do I tell him?" she whispered in his ear.

"The truth; that you are not allowed to be married while you're on the space station."

"But, that's not the truth."

"I know, but he doesn't know that." He winked.

"Um, Kyle, I am very flattered by your proposal, but unfortunately, I cannot marry you."

"Oh, but why?" he sounded like he was about to cry.

"Well, to begin with, I'm much older than you and secondly, I am not allowed to marry while I'm up here. I'm sorry, Kyle."

"Oh, all right." He sounded like he was pouting.

Chandler took the mike; "Cheer up little man. You'll meet a nice girl just like Nadia one day."

"Sorry, Chandler, he went away. I am Christine, and I'd like to ask you a question."

Nadia folded her arms and relaxed in her seat.

"Sure, shoot."

"Do you have a girlfriend back in the US?" She asked with a glimmer of hope in her voice.

Nadia laughed and pointed at Chandler.

He simply winked at her and said: "No, Christine, I don't, and just like Nadia, I am not allowed to date anyone while I'm up here."

"Awwww!"

By that time, the radio signals had faded out. The ISS has only about a ten-minute window for amateur radio contacts.

"So, what do you think?" he asked.

"That was fun." She gave him a sexy smile. "Those children are so adorable!"

"Yes, they are."

"So, what is this amateur radio?"

That opening was all he needed. He began his spiel about his favorite pastime. "It's a two-way radio hobby that millions around the world engage in. You have to take a test and get licensed in your country, and then you can talk with other licensed operators around the world."

"Well, it's delightful. I'll look into it after we get back to Earth. For now, shall we get some dinner?" she asked as she rose like a balloon and made her way to the food storage area.

"Yes, I'm starved."

The astronauts shared the next five radio sessions, and they both thoroughly enjoyed every minute of chatting with those children. They also enjoyed being together and getting to know each other. After only two months of their assignment, they hit it off and had already formed a bond. They had meals together; they watched television together. He read to her every evening to brush up her English skills, and she read to him every morning to help refine his Russian.

One morning, three months into their assignment, Chandler received a notification on his social media account that it was Nadia's birthday. He immediately began preparations to serve cake and ice cream, and even though real candles were prohibited, they had LED candles that could be used. He had no gift to give her, but he did come up with an idea.

While Nadia was sitting in the television area, Chandler and Buddy floated into two of the other seats, holding a cake lit with one LED candle. They both started singing, "Happy birthday to you. Happy birthday to you. Happy birthday dear Nadia. Happy birthday to you!" Nadia's eyes lit up.

"Oh, my! Thank you very much! I never imagined this!" Chandler wasn't sure if there was a glow or tears in Nadia's eyes.

"You're more than welcome," Buddy said as he blew her an air kiss. When he looked at Chandler, he understood that he needed to skedaddle. "I'll leave you kids alone." He floated off to his quarters.

Chandler clasped her hands in his. "Nadia, I have no

gift for you, but I want to tell you how I feel about you. Is that okay?"

Nadia was amazed at how forward and cheeky this American man was. "Yes, that is okay." She was trembling inside.

"Nadia, when I saw you for the first time as we began our training for this mission, I thought you were the most beautiful creature on the face of the Earth. Then, when I got to know you better, I confirmed my thought: you are the most beautiful creature on Earth, both inside and out. I have fallen for you, and I want to be with you forever."

Suspended in mid-air, she didn't try to release herself from his grip. Instead, she held on. They remained suspended, face to face, her hair fanning out from her head in all directions. They looked like two pinwheels, with their arms and legs extended. "What do you mean, 'fallen' for me? I am sorry, but I don't understand this phrase."

"I mean, I have fallen in love with you, and I hope you feel the same way." He looked directly into her eyes and noticed them moistening.

"Oh, love. Yes. I love you, too."

Chandler pulled her close enough so he could kiss her. Initially, his kiss was gentle and brief, but when he looked into her eyes, passion took over, and his second kiss was intense, ardent, and filled with desire. He wanted to make love to her right then and there, but there were rules. No sex – a violation would cause immediate suspension from the position as an astronaut and ended any future with either country's space program.

"Chandler, we must obey the rules."

"Yes, I know we must. But please understand, if I could, I would make love to you right now. Would you want to?"

She kissed him this time. "Yes, if I could, I would, certainly."

"Very well then, let's make a pact."

"What sort of pact?"

"When our assignment is over, we meet somewhere in Europe and make passionate love to each other."

"Can we meet in Paris? I've always dreamed of being in love in Paris." She looked at him with hopeful eyes.

"Paris it is!" he kissed her again.

"This is the best present I have ever received."

***

As the months passed, they became inseparable. They both carried out their respective daily tasks with precision, accuracy, and on time, while kissing, laughing, and fawning over each other every second of their free time. Buddy was quite impressed by their professionalism, especially since he knew well how they felt about each other.

***

The time came for a crew change on the Space Station. As Chandler and Nadia were buckling their belts, they looked at each other.

"Remember our vow," Chandler whispered.

"I will never forget."

***

The return to Earth was uneventful, and everything went according to plan. When their crew cabin came to rest at Cape Canaveral, they completed the tedious checklist that had to be done to make the spacecraft safe. The hatch was opened, and they entered the astronaut transport vans. These were rather luxurious, complete with an extra-wide recliner chair – required because of the bulky spacesuits they wore.

As soon as Nadia was released, two officers came, said something to her in Russian, and took her away. From her reaction, it seemed like she was arrested.

Chandler panicked. "What's going on? What are they doing to her?" He sprinted towards her, but four American Air Force officers stopped him. "Let me go! Let me go!"

Buddy pushed him back by pressing both his hands on his chest; "Sorry, Chandler, you can't go there. Those are Russian lines, and we can't cross them." Buddy was trying to calm him down.

"But they can't do that! She seemed upset. We've got to help her!" He tried to squirm free.

"Dude, calm down, or you'll get us both arrested. You don't want to cause an international incident. That won't help her at all. Now, please sit down, and I'll go talk to her supervisor." He pushed Chandler down onto a bench and went over to the Russian side.

Chandler buried his face in his hands. *This is not good. I've got to do something.* He knew well, however, that if he did anything, he would be arrested. He watched as Buddy and some Russian officer had a civil conversation, but he did notice Buddy becoming agitated as he listened to what the Russian officer was saying. Buddy then turned and went back to Chandler.

"So?"

"I'm afraid I don't have good news."

Chandler shot up. "Tell me!"

"Chandler, sit down please." He waited for Chandler to sit. "The Russians did not take your space romance too keenly."

"Why? We didn't do anything wrong. We just kissed a few times. We got all our tasks completed. You know that."

"I do, but they don't. They're evaluating her story, and if it checks out, they'll let her go. But in the meantime, she's under investigation."

"Good Lord! I can't believe this!" He got up and began to pace. "What do we do? Can we do anything?"

"No, Chandler, I'm sorry. All we can do is wait."

"How long do you think it will take?"

"I don't know. But based on my experience, it may take a while."

All the Russians walked over to a military transport

waiting on the ramp to fly them home. Chandler didn't say a word. He was planning his next move.

As he entered the airport, his cell phone began to "ding" continuously. He couldn't pull it out, though, because he was carrying his baggage out to his pickup truck. As the doors of the terminal slid open, a swarm of journalists, cameramen, and people he didn't know charged towards him.

"Chandler, is it true that you broke the rules up there?"

"Will you be facing consequences for your actions?"

"Will you be seeing Nadia again?"

"Are you going to marry her?"

"Is she pregnant?"

He was taken aback at all these questions, and even though his blood was boiling, he kept his cool. "Get the hell away from me!" He pushed through and quickly made his way to his pickup truck. He tossed his baggage in the back, got in the driver seat, and sped off.

As he reached his apartment building, he noticed that it, too, was swarming with the press. He made a U-turn and headed to his parents' house – a nine-hour-plus drive. He would have plenty of time to think. As he drove, he continually checked his rear-view mirror to make sure he wasn't being followed. He switched on the radio and turned it to his favorite talk-radio channel.

"And now for some curious news: Chandler Barrett, the hot, young astronaut just landed at Cape Canaveral today, and who was waiting for him, there? Of course, the media! They were throwing questions at him, but he simply told them to go away and made his way through the crowd, got into his truck, and took off. He seemed quite annoyed. No one knows exactly why. He should be happy that he has captured the heart of one of the most attractive female astronauts on Earth – or in space! But Mother Russia was not going to stand by and let some obnoxious American sweep their darling off her feet! Oh no. They made her disappear quite quickly. Will he ever see her again? Will Nadia be locked up? Let not your heart be troubled; we'll follow this story very closely!"

"Obnoxious? Really?" He sped up as his blood pressure rose. He changed to his favorite rock channel. He kept it there until he reached Buford, where his parents lived. Since they had listened to a lot of his favorite music in space, all the songs that played reminded him of Nadia. A plethora of emotions bombarded him. He was angry. Frustrated. In love. Depressed. Hopeful. Hopeless. It was a wild roller coaster with no operator. And he was in the front seat!

Finally, he arrived at his parents' house. His father met him in the driveway.

"Welcome home son!" He gave him a big hug.

"Hey, Papa, how you doin' Old Guy?" His southern Georgia accent came right back.

"Well, bless your heart! You came home!" His mother walked out of the house, wrapped her arms around his neck, and kissed him on the cheek several times.

"Yes, Mama, I came home… you can let go now."

"Oh, I see. The big astronaut is too big to hug his Mama!" She laughed and gave him a big, loud smooch on his cheek!

"Ma!"

He put his arm around his mom and they both walked into the house while his father took his bags from the truck.

"Oh, Channy, what happened up there?"

"Mama, when are you gonna stop callin' me Channy?"

"Never! You're my baby boy, and I luv ya!" She pressed her lips on his forehead and gave him another very loud kiss.

"Dagnabbit, Terry! You leave that boy alone. Can't ya see he's distressed?"

"Fine! I'll pour you some sweet tea while you tell us what happened up in space."

Chandler took a deep breath and told his parents exactly what happened during those six months.

"You never touched her?"

"No, Papa, I swear! Besides, it's forbidden."

"Well, that famous apple was forbidden to Adam and Eve, but that didn't stop the devil from temptin'."

"Yes, I know, Mama. But it's very difficult up there to do even the simplest task, let alone hanky-panky!"

"So, what now, son?"

"Well, I gotta keep a low profile for a few days. Then

I'm gonna contact Buddy. He was the mission commander. He saw everything, and he can be a witness for both of us."

"Witness? Is there gonna be a trial?"

"Oh, no, I hope not." He pulled his cell phone out and dialed Buddy.

"Chandler, what's up?"

"Buddy, I'm at a loss here. What do I do?"

"Where are you now?"

"At my parents' home."

"Good. Stay there until you hear from me, okay?"

"Yes, I can do that, but I want to talk to her. I need to tell her how I feel."

"Chandler, she knows how you feel. Now, listen to me, don't contact her in any way, shape, or form. Do you understand? You could get her in deeper trouble. I'll work with the Russian government and explain the entire ordeal. But you stay put. Got it?"

He frowned as he nodded. "Got it. But for how long?"

"I'm not sure. I'll check in with you in a few days."

Over the next few days, Chandler searched for Nadia everywhere: on all the social media sites, email, and the web, but she seemed to have fallen off the face of the Earth.

"Why don't you try writing her?" his mother suggested.

"Write her? You mean a letter?"

"Yes, I mean a letter. And mail it using a stamp. Who knows, she may actually get it."

"A letter, huh..."

"Here!" His mother dropped a large notepad with a pen in front of his face.

He picked it up and kissed his mother. "Thanks, Mama, I owe you one."

"You owe me plenty, Channy!"

He rolled his eyes and picked up the pen. He scratched his head and began to write.

My Dearest Nadia,

I was mortified to see them take you away like they did. I didn't know what to do. I came to visit my parents, and my mother suggested I write to you, so here I am, writing to you.

I'm going to keep my side of the promise and meet you in Paris. I will see you on May 16, 2021, under the Eiffel Tower at 10:00 a.m. sharp. I pray that you will be there to meet me. If you come, I will have a surprise for you.

Please don't stop thinking of me, because I can't stop thinking of you.

I love all of you with all of me.

Your forever love,

Chandler

"There! Would you read it, please?" He gave the paper to his mother.

She put her glasses on and read it. As she read the

letter, she smiled, and her eyes were smiling, too. "Lovely! Just lovely!"

"You really think so?"

"Yes. So, what's the surprise?"

"You'll see."

"Oh, Channy, I already know, silly. I'm your Mama, ain't I?"

***

Chandler was standing under the Eiffel Tower looking at his watch. It was five minutes after ten. Each minute seemed like an eternity to him, and all sorts of thoughts were going through his head. He looked at each and every person who passed him by or was in his vicinity. Finally, he saw her on the south side of the tower. She had cut her long blond hair and was wearing a flowing, floral dress with white sandals and a pink sweater. He bolted towards her, and she did the same. As they approached each other, their arms opened wide. They hugged, kissed, and twirled around a few times. Many of the passersby stopped to see what was next, and a few of them pulled their phones out just in case there was some video recording to be done. They did not disappoint. They finally stopped, and while they remained embraced, Chandler said, "Oh, I'm so glad you're here! I thought I'd lost you forever. I was having all sorts of weird thoughts."

"Me too, Chandler."

He kissed her again but didn't let go. "How did it go with your government?"

"Good. Buddy came to my rescue. He gave them a thorough report on what happened up there, and especially highlighted the fact that we were not intimate."

"I knew he would come through! Thank you, Buddy! Wherever you are!"

She giggled. "He can't hear you."

"I know, but I will thank him in person."

Nadia turned and looked around. "I think we have company."

Chandler also noticed many spectators with their cell phones in hand. They must have recognized the couple. "Well, I suppose we ought to give them what they're waiting for."

"What?"

Chandler pulled a small box out of his pocket, opened it, and got down on one knee: "This. This is what they're waiting for. Nadia Katarina Subciek, will you marry me?" He really hoped for a positive answer. *If she says no, I'm going to look awfully foolish!*

Nadia started to tear up. "Yes, Chandler Robert Barrett, I will marry you." She got down on one knee and kissed him.

He lifted her up. "You're not supposed to get on your knee; only I am to do that." He laughed.

"Well, I'm sorry, I've never been proposed to." She laughed too, along with the crowd who cheered, clapped, and laughed.

# # #

Joanne Fisher is a Canadian-Italian-American author of the book "With All of Me" and the short story "The Christmas Concert," which appeared in the SCWG's anthology *Holidays*. She is president of the Space Coast Writers' Guild. She is married to the love of her life, Dan, and lives in central Florida. Follow her writing on Facebook @ReadJoannesBooks, on Twitter @JoannesBooks, on Instagram joannes_books_2018, or online at www.joannesbooks.com.

The International Space Station photographed by an STS-134 crew member on the Space Shuttle Endeavour after the station and shuttle began their post-undocking relative separation. Undocking of the two spacecraft occurred on May 29, 2011. (NASA)

It's human nature to stretch, to go, to see, to understand. Exploration is not a choice, really; it's an imperative.

Michael Collins, Gemini and Apollo astronaut

SPACE: 1999

Scott Tilley

In 1975, *Star Trek* was gone from the airwaves and hadn't yet returned to the big screen. *Star Wars* was just a gleam in George Lucas' eyes; *Episode IV: A New Hope* was still two years away from its premiere. Even the corny *Buck Rogers in the 25th Century* wasn't released until 1979. For science fiction aficionados and space junkies like me, it was a pretty barren time.

Space: 1999 opening from Season 1 (Wikimedia)

Into the void stepped a quirky British television series called *Space: 1999*, which made its broadcast debut over forty years ago. For a young boy eager to experience everything about space, watching the show on Saturday night became a tradition. I first saw the French version in Canada, called *Cosmos: 1999*, and later saw reruns of the original English version.

Produced by Gerry and Sylvia Anderson, *Space: 1999* was an interesting idea about an international group of scientists stationed on Alpha, a permanent base on the moon. When nuclear waste stored on the moon explodes, it acts as a gigantic rocket engine and tears the moon out of Earth's orbit. All the stories that followed the first episode, "Breakaway," used this catastrophic event to structure the show's narrative.

The spaceships used on Alpha were called Eagles. They looked rather like a skeletal version of the Space Shuttle. I was quite enamored with the Eagles: I had a plastic model that I built and hung from my bedroom ceiling. (It was later joined by an X-Wing fighter.)

Eagle on Moonbase Alpha. (wallpapersafari.com)

From a viewer's 1975 perspective, 1999 was over twenty years in the future – long enough to be optimistic about exciting new technology, crewed space travel, and alien adventures. For example, all characters used a CommLock to communicate. This device looked somewhat like a remote control with a small TV screen on the top. It was also used to open doors and control other equipment. The CommLock was a reasonably accurate vision of today's smartphones.

The crew sometimes carried portable laser weapons, called Stun Guns. I was so taken with them that I actually owned a plastic toy version. Today I own several real lasers, but I use them for much more peaceful purposes: to play with my cats.

Season 2 cast, metamoprh Maya center (Wikimedia)

Space: 1999's premise that the environmental dangers of nuclear waste would lead to humanity's downfall was prescient. When the moon left Earth's orbit, the planet's inhabitants were devastated due to massive gravitational forces that caused earthquakes, storms, and other natural but extra-powerful disasters.

Unfortunately, we never returned to the moon. But hopefully, someday we will. How else will we meet metamorphs like Maya, androids, and sentient plants?

# # #

Scott Tilley is an emeritus professor at the Florida Institute of Technology, president and founder of the Center for Technology & Society, president and co-founder of Big Data Florida, past president of INCOSE Space Coast, and a Space Coast Writers' Guild Fellow. His recent books include "Systems Analysis & Design" (Cengage, 2020), "Technical Justice" (CTS Press, 2019), and "Dreams" (Anthology Alliance, 2018). He wrote the "Technology Today" column for FLORIDA TODAY from 2010 to 2018. He holds a Ph.D. in computer science from the University of Victoria. Visit his author website at http://www.amazon.com/author/stilley.

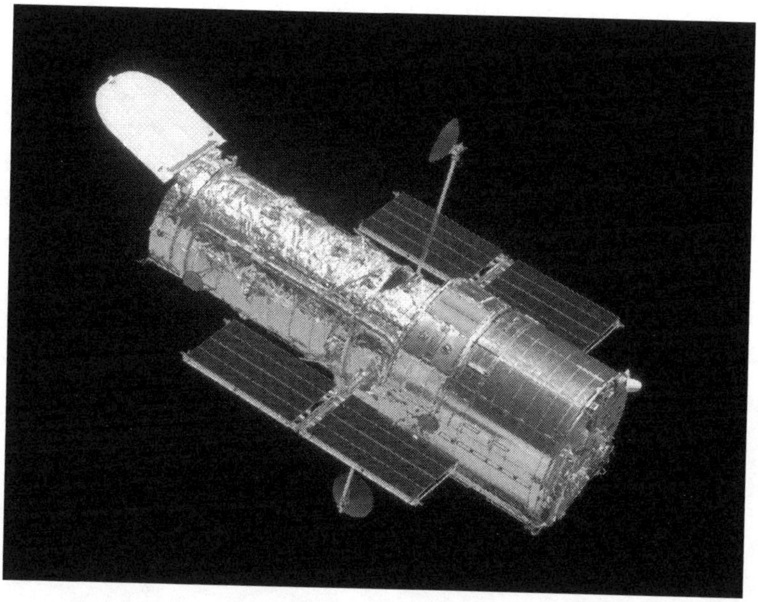

The Hubble Space Telescope as seen from the departing Space Shuttle Atlantis, flying STS-125, on May 19, 2009. (NASA/Ruffnax)

There is nothing so far removed from us to be beyond our reach, or so far hidden that we cannot discover it.

Rene Descartes

Merritt Island National Wildlife Refuge Family Excursion

Betty Whitaker Jackson

What a privilege it is to live on the Space Coast of Florida in Brevard County and to experience all the benefits the space program has brought to our region. These attractions have allowed our family to expand its horizons, to enhance our children's education, and to know many professionals involved with NASA industries. With get to see, hear, and feel the rumbles of spacecraft launched into the skies. Even our area code is related to the space program: 3-2-1. Liftoff!

We have lived in this area twice phone , first in the 1980's and 90's, and again from 2007 to the present. We have witnessed live launches, heard astronauts speak, followed lectures and informational broadcasts on the NASA channel, visited the NASA complex numerous times, and participated in Space Odyssey musical events with the Space Coast Symphony Orchestra. We have heard narrations by Senator Bill Nelson and several astronauts as part of those performances.

Our son participated in Space Camp and has never forgotten the experience. We've collected patches from flights, eaten space ice cream, toured NASA's facilities often, taken our tourist friends there and watched their amazement as they saw the rocket park and Space Shuttle Atlantis. We even know of a wedding held in that facility where the IMAX film entertained guests. We are truly grateful to live here.

***

Our most memorable recent event was loading our son's seven-passenger van and touring the Merritt Island National Wildlife Refuge (MINWR) with our whole family. My husband and I have gone there each February to see the migratory birds that populate this 140,000-acre site surrounding and protecting the actual launch pad and other facilities. It is a buffer zone for launches that has been preserved as a wildlife habitat of unbelievable beauty and diversity. It is open to the public (except when launches are imminent) and is an absolutely delightful location.

We take photographs of animals, birds, wildflowers, and wide-open expanses preserved for us to enjoy. The fifteen-mile dirt road meanders around the area, and frequent pullovers allow photographers to come from all over the world to enjoy what nature provides.

On this particular day, our granddaughters stared wide-eyed, holding their tiny kid-cameras, urging their dad to "Stop, Daddy!" every twenty seconds or so. They recorded their visit in snapshots and videos, and their five and seven-year-old senses of "Wow!" erupted in squeals of delight.

Not only were the children witness to amazing scenes, but our son, daughter, and daughter-in-love also delighted in the peaceful, slow meandering drive, and were transfixed at sights of birds, flora, and fauna around every bend. It was such joy for my husband and me as well, because he's usually the driver, trying to take in all the scenes while driving. Today we had a chauffeur, and our son was most patient with requests to "Slow down, Daddy. I think I see something!"

It became a game of "I Spy!" not only for hidden wildlife but for a soon long-list of license plates from all over the country. The catalog expanded from the usual Florida ones to fourteen other states' representatives. And, probably for the first time, the girls saw cameras with long lenses, obviously able to focus on the spoonbills nesting almost out of our sight, mere specks of pink to them. (Of course, their favorite color.)

Until we had taken this trip, our Kaley thought all ducks were mallards. She already knows the males are colorful, and the females are dull-feathered in the bird world. As a budding scientist, she was amazed at the variety of ducks swimming in this preserve. And she had to tell us, with great authority, of course, because as a seven-year-old, she knows for sure, that each bird had a different kind of beak so it could eat different kinds of food. We listened as she taught these facts to her little sister, who couldn't contain her excitement at spotting birds hiding in bushes or mangroves.

Their greatest delight was in spotting the alligators. I knew where the granddaddy was, on the exit road, so I told them repeatedly, "Wait 'til you see him!" Meanwhile, the children cataloged no fewer than seven others they spotted, to their utter delight (and some trepidation). Some were swimming; one was stalking a duck. Oh, the shouts we heard. "Watch out, little duck. He's coming!" Fortunately, we did not witness a grizzly encounter. Some of these reptiles were sunning themselves on hammock grasses, their armored tails quietly baking in the afternoon sun.

***

When the chance came, the children took the walking path out into one of the viewing areas. There, they hoped to see something different from the things they had already seen while touring from the van. Alas, it was not to be. When I asked the eternal question, "Well, what did you see?" Kaley answered, "Nothing special, Grandma."

I was immediately reminded of Helen Keller's delightful essay, "Three Days to See," a piece I taught in school several times. In it, she imagines what she would love to see if she were given the gift of sight for just three days. She admonished a friend who had just returned from a walk. She asked that same question and received a similar answer. In chagrin, she wrote how that could never be. Her hands felt the outline of every leaf and the roughness of the bark of a tree. She knew that if she had the gift of sight, nothing, absolutely nothing, would be ordinary.

I think of that essay often, since I've just been through the frightening experience of limited vision after a retina episode. Never again will I take sight for granted, and I so wish I could see with the perfect vision my grandchildren have. I am teaching them to observe carefully and to build sight memories.

***

Last year, MINWR was flooded, and there were talks they could not repair the roadway. In the last few years, it has been devastated by wildfires and drought. However, this wide expanse of land is a living thing, and although humans has left little imprint on it except the roadway and obvious drainpipes and other water-control measures, it has been allowed to ebb and flow as nature does. I know there must be some management, but it is good to know that when a tree falls, it is not immediately removed, as if it never occupied the space. It is allowed to provide shelter, to grow green moss, to provide bas relief for its surroundings, and to just be there. I delight in the gnarled bark and limbs of the old cedars that have been there probably as long as my seventy-eight years. The stories they could tell.

Perhaps the best thing about our visit that afternoon is that the experience has stayed with the children. They keep asking to go back. Out of the blue, as we're at the beach, or watching things in the garden grow, or seeing a butterfly, they will remark, "Grandma, remember we saw these butterflies at the Space Center?" I've been rewarded in knowing that our invitation to go see the Wildlife Refuge was a good one. We experienced so much, but mostly, it represented a together time of adventure, a recognition that God's world is beautiful, and that we need to preserve such places so someday they can drive their children through the Merritt Island National Wildlife Refuge. I pray they will.

# # #

Betty Whitaker Jackson has written nineteen books since her retirement from her teaching career. She is an active member of Space Coast Writers' Guild, has work published in eight anthologies, and won First Prize in the Life-Rich Reader's Digest Memoir Contest in 2014. Her books are available in paperback and ebook formats at bettyjackson.net. Her favorite writing quote is "I write to discover what I know," by Flannery O'Connor.

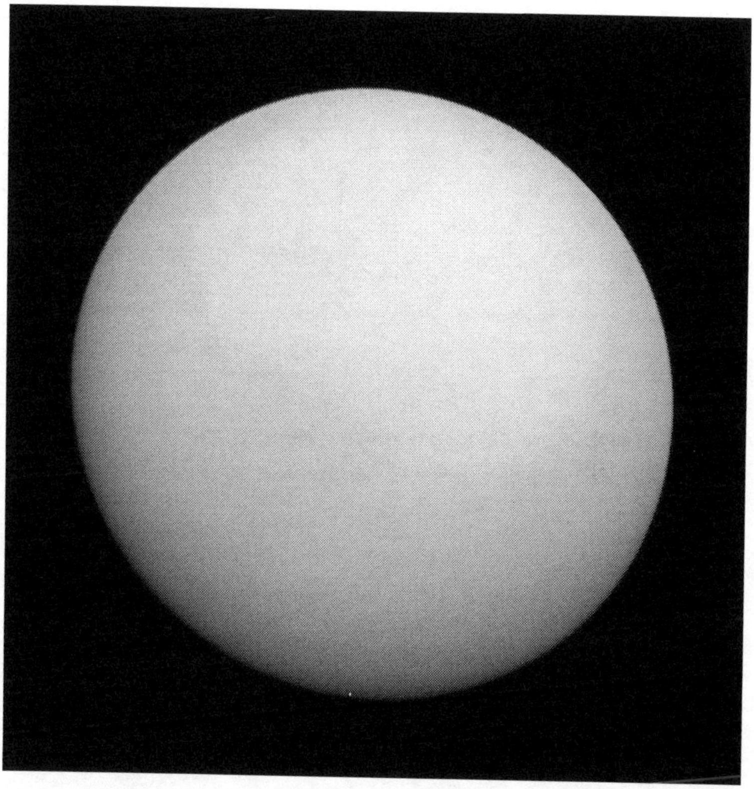

Uranus, taken by Voyager 2 in 1986. (NASA/JPL)

Why our space program? Why, indeed, did we trouble to look past the next mountain? Our prime obligation to ourselves is to make the unknown known. We are on a journey to keep an appointment with whatever we are.

Gene Roddenberry, Executive Producer of "Star Trek"

VALLIS

Richard Marschall

Prologue

On the Martian landscape, Peace Vallis is thought to be the dried riverbed of a once magnificent surface river that meandered gracefully for hundreds of miles through what is conjectured to have been a beautiful, verdant plain. The river was bordered on all sides with vibrant vegetation and sported a thriving population of hundreds of thousands of the humanoid bipeds. Their remains were found on the third Martian exploratory mission in 2035.

It is 2045, 0500 nautical time.

Phobos, Mars' largest moon, glared down on the small planet from its orbit nearly six thousand miles above the surface. Three billion years before, the Great Cataclysm, as humans would call it, left its indelible impression on the surface of the solar system's fourth planet from the sun.

The gods ruled Mars from their perch high atop Olympus Mons, and King Arwyn and his family governed the lands and the people. Mars had been a marvelous planet of flowing rivers, productive soils, and an atmosphere not unlike that of her sister planet, Earth. Arwyn ruled justly, and under his leadership, the enclaves flourished, and his people were happy.

Unknowingly, there were dark forces brewing underfoot that would be unleashed once the great catastrophe struck that proved far greater than any one man or person.

***

The winds were beginning to whip up the dusk on the dried Martian landscape – reds, browns, and strange shades of oranges blended in a maelstrom of colors. Beneath the surface, the waters moved swiftly, and the subterranean fishermen made their catches for the day - white crusted amphibious things with bulging black eyes and spikes on their skins. They were the delicacies of the Mortmans, the people who called the cavernous underground their home.

***

In the dark corridor of space between Earth's Pioneer Space Station II and Mars, a spaceship is on route to the red planet to build the first permanent one-hundred-acre colony, which will be capable of eventually housing over 100,000 individuals fleeing the dying Earth. These are the lucky individuals fortunate enough to win the great drawings taking place each month in the polluted, oxygen-deprived environs of Terra Firma. Scientists envision two to five million people living on the Martian landscape by the year 2060.

***

Beneath the dried Martian landscape, millions of beings whom we might call heinous by our standards live and thrive in the underground corridors and caverns. Hot springs from the planet's core fuel the energy needs of these creatures and provide warmth, light, and a place to dispose of their wastes. Far below the strata where these beings make their home, the temperatures of these springs exceed 2,500 degrees.

The Martians are fortunate their planet has an abundance of minerals, such as iron, nickel, copper, magnesium, aluminum, calcium, and potassium within accessible reach. Large deposits of phosphorus provide a source of light for the otherwise black interior.

***

The Mortmans have lived here for millennia; no one knows exactly how long. Recorded history dates back to only three billion years ago when the Great Cataclysm took place. Conduits from the searing springs feed furnaces where the frozen underground rivers are melted and distilled. The underground caverns are endless, and even today, millions of square miles remain unexplored.

The Mortmans are unlike any creature on Earth, human or animal. Intelligent, with sharp, arm-like appendages similar to those of the terrestrial Tyrannosaurus Rex with three fingers that can be used to manipulate objects and equipment. Their heavily-muscled, darkish-green legs are used to propel them. Their heads resemble a cross between fine-cut porcelain doll with large, yellowish eyes and the exquisite look of an ancient Egyptian carving. Their movement is fast, almost staccato style, with a gait like that of an iguana, their bow legs moving fast, then slow.

"Ignaeous, have you completed the new archive chamber yet?" the smaller, obviously female Librarian inquired in an almost lyrical voice.

Concluding his seemingly frenetic placement of the silver-colored disks into the grooved tracks of the brand-new bookshelf, the older, white tousled-haired figure responded. "Monaea, when is the last time you have been to the surface?"

"Two days ago, *why*?" inquired the attractive, black-haired lady attired in the tight-fitting bluish-green outfit.

"Something strange is going on in Level Eight. The illumeens have been flickering, even going off and on," he replied. "I was wondering if there are storms on the surface."

"*No!* There was nothing unusual when I was up there," Monaea responded.

Sitting down, cupping his head in his three-digited hands, he let out a heavy sigh before he went on. "Monaea, I haven't told anyone else, but I have been experiencing strange sensations recently. During nocturneel, it is like there are other presences here in the Library – presences that can't be explained."

"Enlighten me, my superior. I am not comprehending what you are telling me."

"If any of this ever leaks out, the Council will send me to the Archipelago and throw away the key," the older man said reluctantly.

"Ignaeous, you are scaring me! *Please* stop!" his assistant pleaded.

"I cannot! Already I have told you way too much.

Please, *allow* me to continue."

"I don't want to hear anymore! They will think me complicit in your ramblings," Monaea replied.

Staring into space, absorbed in his thoughts, the old man continued anyway. "Four weeks ago, I heard a noise, like air being sucked into the compressor, only it wasn't air, and the compressor wasn't running. When I entered the Reference Department, I noticed that dozens of books had been strewn on the floor. I thought maybe the children who had been there earlier in the day had wantonly left them on the floor for the old man to clean up. No problem! But when I went to pick up the books, one by one they began to mysteriously open, displaying pages of one Mortman family history. It was the history of the Ktroio-Riehke family," the older man slowly revealed.

Now more enthralled in what transpired, the younger librarian asked, "What did you do?"

A loud rumbling and an enormous vibration shook the chamber in which they were standing, stopping their conversation. It was as if someone had struck the top of the room with a giant sledgehammer, a sound so deafening it caused them to lose consciousness. When they came to, they were on the floor, the young woman atop the old Librarian. "Monaea, talk to me," was all the older man could think to say.

"Ohooooo, my head hurts," the young woman said as she grabbed her head with her hands.

"We've been impacted again," the older man responded.

"That's the fourth time this month!" Monaea replied.

"We can be thankful we live *beneath* the surface," the older man shot back. "This must have been a big one, probably one of the asteroids. It's that season again!"

Getting up and checking the room for damage, they saw that many of the bookcases and displays had been overturned. It would take many days, perhaps weeks, to repair the damage. They had forgotten what is was they were talking about before the impact.

***

Away from the Library, down the dimly-lit corridors running adjacent to the Humanity Center, there were essences at work. They were scouring the walls, looking for signs, inscriptions, and any indications of past references of the Ktroio-Riehke family. These records would shed light on their mysterious disappearance three billion years ago.

At first, they were just flickers and flashes of bluish-green light. As moments waned into minutes, shapes and figures began to emerge. Vapors coalesced into solid forms – seven figures intent on learning how and why their forebears were extinguished and stricken from the records of Mortman civilization. Their yellowish eyes glowed a menacing red. They searched each niche of the subterranean cavern that had been the home to the Ktroio-Riehke family. When their search bore no fruit, they assembled over the old stone slab that had once been their family's centerpiece and made a silent vow that not until their quest had been satisfied would they leave the others alone. Calling up the Spirit Force from below, a great rush of fiery lava engulfed the room, and a headless presence spoke out to them. "I command you to retake your positions as the rightful rulers of the Empire. Only then will you be free of the curse set upon you by the ones who slew you and your family so many years ago. *Slay* all who oppose you!"

The voice and the fiery lava were gone, back to the depths from which it had come.

***

Having completed his inspection and inventory, Ignaeous was astounded to see that all the books on the Ktroio-Riehke family had been neatly replaced on the shelf where they belonged. "Monaea, this is unreal! How, in the aftermath of the impact, could these books have returned to the shelf?"

"Ignaeous, are you certain you didn't replace them yourself before the incident?" asked Monaea.

"Darling, I know you think I'm clumsy and dimwitted at times, but I'm *not* crazy," the older man retorted. Then, smiling, he added, "Perhaps I *am* losing it."

Monaea looked at him quizzically, tensing for what she believed might be a breakdown of some sort. Then, seeing the thin smile spreading across his face, she realized he was jesting, and she relaxed. "Ignaeous, don't do that! You know I love you. We have worked together for too many years for me not to believe you."

The rest of that day they worked in silence. By day's

end, most of the books and debris had been cleaned up, and both Ignaeous and Monaea were very tired. At 9:00 p.m. the lights dimmed automatically, signaling shutdown for the evening. "Goodnight," they bid each other as they made their way to their respective quarters.

***

In the dimly lit corridors, the specters made their way noiselessly from the walls of the Humanity Center to the more populated areas of the Hroletariat Common, where most of the populace lived. They had been instructed to spare no one in their quest to re-usurp their rightful roles as rulers of this world. This admonition weighed heavily on the old king.

They were drawn, like bears to honey, to the diaries and documentation manuals in the Library, for it was here they might glean the truth of what happened to their family.

They floated silently through the walls, before allowing themselves to become humanoid presences. They had only a short time before they would again return to their more ethereal forms, so work would have to be done quickly.

Dropping all the family books and manuals that had been so well preserved in the Martian underground to the floor, they began reading in earnest. It seemed that many eons ago, the ruler of Rubicund, as the planet was called then, was a benevolent, caring, and just man by the name of Arwyn. The planet flourished under an atmosphere of enriched oxygen, and the people prospered under Arwyn's rule. For many millions of years, the red planet produced an abundance of crops, and the people were happy and content. But then, inexplicably in the year 3225 10th BC (before the Great Cataclysm), everything changed.

***

The world that had been, one of harmony and grace, ceased almost the moment the tremors shook the ground, and the great cities caved in amongst themselves. Entire civilizations vanished, and the seas retook the lands. The great king was replaced by bands of roving marauders who plundered the once-great manors and estates. Harmony gave way to discord and fear. When the king and his family returned from their retreat in the hills, there was nothing left, no homes, no factories, no food. Families disintegrated into bands of roving migrants, searching for food and sustenance. Only the stronger clans survived, but at the expense of others. No longer would they accept the dominance of a king. History was changed in a heartbeat.

Eventually, the rivers drained into the earth, and the surface lands became barren and dry, and the oxygen dissipated into the atmosphere. Everyone that could went underground to start a new existence. These were the new Martians, the ones who would subsist beneath the surface.

The Kitroio-Riehke family, the clan of the king, retreated from their home high atop Olympus Mons. Theirs was a big family with influence reaching the far ends of the known world. Strongest in Vallis, it was here they made their stand. Six thousand strong, they began the arduous process of retaking the lands. It was hard, and there were many bitter battles. Even when they had reclaimed a large part of their original kingdom, they found the soil was no longer workable, the winters longer, and the air more difficult to breathe. They took their family and subjects with them, deep into the underground caverns close to the great subterranean rivers lined with the phosphorus that glowed in the dark.

It was difficult at first, but they soon learned to fish the untainted waters and prepare the strange-looking fish for their meals. They discovered there were plants they could grow and harvest in the semi-darkness, and they found minerals to rebuild their cities. Soon many additional thousands were living on the lands beneath the red surface.

But all was not well. Resources were scarce, and space was limited. Even though King Arwyn and his Council worked hard to provide, it wasn't long before fights broke out among neighbors, which escalated into disputes in the General Council. Discord was rampant in the colonies. "Why is it the gods are fat and want for nothing, while we mortals must struggle each day for our survival?" the people asked.

Infighting and disagreement within the various wards of the underground kingdom became common, and a few years later a new leader emerged, a leader who would soon enjoy the backing of the Hroletariat, whose numbers by this time far surpassed the members of the royal family.

***

A night that no one would remember (other than a few dozen black-clad mercenaries) would go down in Martian history as "The New Horizon." All records of the Ktroio-Riehke Family were erased from the new Martian landscape below the surface. No one would ever know of the atrocities that took place that night. They would not know of the horrible, senseless, degrading tortures that occurred, and how the dismembered bodies of the king and the six highest-ranking members of the Kitroio-Riehke family had been thrown into the fiery pits below.

A generation later, no one even remembered the once-great king and his benevolence and grace, his humble demeanor, and his strength. All that remained was a dark cloud that, after a few more years, ceased to exist altogether. The new regime sealed all the records of the deposed family into the walls of the new Great Humanity Center, which by then was the newest sector of the caverns. By law, it was decreed that no one could ever speak of the family again.

The family faded into oblivion, and the deceased suffered the pangs of their torture and subsequent disposal. Their material bodies gone, their spirits were sealed into the hard, granite walls of the Martian underground. They would not be heard from again for nearly three billion years

***

Fast forward three billion-plus years to the present. A technologically advanced civilization had expanded its below-surface population, and there were now millions of people living in the underground corridors. Life was good, but something was missing.

With all their material needs satisfied, the people had turned to more leisurely pursuits. Now a socialist-populist democracy on the verge of collapse, the morals of the people's representatives had grown corrupt and centered more on their own needs and re-election than on the needs of their constituents. Politicians openly sold their votes to the highest bidder. Unethical legislators did not hesitate to wantonly tax the rich and distribute their hard-earned money to the poor.

This newly-empowered group soon gave rise to a new perspective. No longer faced with the need to be productive members of society, their only responsibility was to make certain they voted for the candidates who promised them the most benefits. Their leaders promised not only a continuation of these unsustainable benefits to the detriment of the more productive members of Martian society, but an escalation of new programs that would give the lower classes even more.

***

They had been erased from history. Their physical bodies burned from their bones and disintegrated in the fiery abyss; their souls locked in the solid walls of the chamber where they had suffered their horrible deaths. Embedded in the solid granite walls, their life essences were stuck in limbo: neither the Martian landscape nor something beyond.

All seven felt the rumbling underground and marveled at their release. How easy it was for them to float through the corridors and enclaves of the modern cities. It was strange how they floated undetected by others they passed on their way.

***

Descendants of the once-great ruling family were scattered over the far reaches of the new empire. Stripped of any power or authority, there were none with even a memory of their once-great ancestral forebears who had lived and ruled the surface long ago. It was gut-wrenching. It tore at the very hearts of the seven departed rulers. King Arwyn believed it was his duty, his destiny, to make right that which had occurred so long ago.

"Arwyn, how are we to change that which has already taken place?"

"Jati, the answer lies in the powers granted us by the Firepeople."

"My highness, I don't understand."

"You will, my friend. You will!"

The rush of air from the vents high in the cavernous ceiling made a whooshing sound as it exited the big vent. Arwyn unconsciously braced himself, although he knew that in his ethereal state he would be unaffected by the rush of air from the massive fan in the giant conduit above.

"Arwyn, *must* we slay all our opponents?" Jati asked, looking down at where his feet should have been.

"Jati, the Firegods have spoken. There is no other way. They will materialize us long enough to accomplish our mission. Then we can pass along our hierarchy to our heirs."

"But, can't our mission be achieved without so much bloodshed and killing? Too many have suffered already!"

"I'm afraid not! Violence begets violence, and only then can peace ensue," Arwyn responded, a bit annoyed his senior cohort and brother-in-law would be questioning his plans.

"But my liege, *why* would the Firegods bestow upon us such awesome temporary power? What is it they have to gain from our ascendency?" Jati asked submissively.

"Why, I *don't* know! I assumed they were on our side. But possibly that is a false assumption," Arwyn replied. "Is it not written that we show best by example? Is it not in our best interests to bring our people together? A short insurrection, then a smooth, gentle transition. It is not unheard of that resurrections are often accompanied by bloodshed and death."

"My king, you have always ruled by your grace and compassion. It was not because you were a poor leader that you were deposed; it was because you believed in the goodness of others. You did not see the deceit and treachery your opponents planned until it was too late. Do *not* become like them," Jati pleaded.

"Your point is well taken my friend, but I must think how we get the populace involved again, how we awaken them from their complacency," the great king responded.

"That is what you are best at my King." Jati smiled.

***

The corridors were quiet at this time of night. Time under the surface was artificial; clocks were set to the hours of daylight above. Morning came at 7:00 a.m., and lights were dimmed at 9:00 p.m.

In the darkness, the seven spirits worked feverishly to locate the missing bits of information that would allow the people to see the fallacy of their ways. They would have to utilize the brief period the Firegods would give them to reappear as mortal beings. Whatever they did would have to be done with extreme precision.

The balance of history could not be changed too drastically. It would alter the time continuum and irrevocably change the order of history. It could drive the entire planet into a space warp and possibly into another, unknown dimension – perhaps even killing off the millions of people they wished to save. They would have to work carefully yet purposefully.

***

Their dark robes hung listlessly in the Hroletariat Common. Each specter understood their goal. On the large screens hanging on the walls, the news of the day played out monotonously as passersby greeted each other with the raised fist and stomp of their right foot. The one called Fallon, who seemed to instinctively understand the pulses of electronic circuitry, went at once to the large steel panel behind the forum and removed the shiny white cover. In seconds, he had manipulated the switches to where he now had control of what was shown on the vast screens. "Arwyn, move to the small lens where everyone can see and hear you."

Appearing in his black robe with the sequined green borders, Arwyn closed his eyes a moment before he began his impassioned plea to the Martian population. He knew in his mind what he had to do. The resurrection would be quick, ruthless, and would spare no one even remotely involved in their demise three billion years ago. *War was hell, and there would be penalties to pay.* His family had been slaughtered. The perpetrators must be held responsible, their heirs held accountable. They must suffer a similar death.

"My fellow brethren, you do not know me, which is of no consequence, but I do know you, and it is most urgent you hear what I have to say," Arwyn began.

Carefully phrasing his words, the tears welling up in his blue-grey eyes, the old leader continued. "Recently (he lied), I'd learned that the history of one ancient Martian family had been stricken from the records, relegated to the Negeons of Dark Information. These records have been hidden from the people for many, many years. Only the Librarians and university researchers, who no longer even care, had access to these ancient archives." Arwyn stopped to look around the great Hroletariat and assess the demeanor of those in attendance.

"*Why*, you ask, is this information of any relevance to you?" Stopping a second time, he saw the blank, indifferent expressions on their faces. He was not getting through. Looking out the large two-way window from the broadcast cubicle in which he was standing, he saw the younger children giggling and squirming by their parent's sides with looks of strained, childish boredom written on their little faces. He would need another tact.

Striking a new, more forceful tone, the old leader continued. "The Kitroio-Riehke family once ruled the nations of Mars. They were good rulers and put the welfare of their subjects first. Under their rule, the people prospered, and the economy flourished. Everyone was happy and content, looking forward to long lives of fulfillment. Then, ..." Stomping his foot and slamming his fist onto the table for effect, "it all changed!"

"What happened?" yelled a petite lady with yellow-green hair from the far side of the common.

"How'd it change?" shouted a short, overweight man wearing wide purple suspenders in the front of the audience.

Arwyn replied. "It was strange! A great tremor shook their home, and an enormous wave of earth sped across the land, buckling roads, buildings, and entire cities that had taken eons to build. Qtillions of dollars of property were wiped away in minutes. Chasms opened in the earth that swallowed the hard work of an eternity. Civilization as we knew it was no more. All the king's resources were not enough to restore that which had been lost, and soon the benevolent king's empire began to disintegrate. Cities lost contact with one another. The once magnificent Martian surface began to change. The surface waters sunk into the earth, and all plants and foliage began to wither and die. Within months, the surface was no longer livable. The atmosphere that sustained the Martian people thinned; nights became longer and colder. The great plains no longer supported the grains and vegetables the people needed to survive. In time, there was not enough food to feed everyone. Even the king's vast storehouses were depleted."

"How do you explain the riches we enjoy today?" a raven-haired lady called out from the back.

"Aaaaah! But the king was still wise, and he took as many of his followers as would go with him beneath the surface. You see, in their final days on the surface, the king's scouts and scientists had discovered there were vast underground caverns, many of which had been newly opened by the Great Cataclysm. Vast rivers had been formed when the waters were diverted from the surface. Fissures opened by the great earthquake radiated heat from far below the surface, and natural phosphorescence provided light for the weary people fleeing the surface," the old king related.

"Life was good beneath the surface, and the people again thrived. Many years passed, and the king grew very old. Others created dissent within the underground kingdom and persuaded the people that their lives left a lot to be desired. False promises of greater wealth and government largesse weakened the king's power. The king was old and perhaps did not see the maelstrom boiling below the surface. He thought he could contain his subject's disillusionment, so he held his ministers in check, not allowing them to take remedial actions."

"Then, it happened, the unexpected! In an overnight coup, the dissenters cruelly butchered the royal family, the king, and his six closest ministers, all members of the Kitroio-Riehke family. They destroyed everything remotely related to the family, even the records and daily journals. Only a few documents related to the government were salvaged to be sent to the farthest, deepest archival tunnels. Nothing was ever heard of the royal family again. Eventually, it was made unlawful to even speak of them."

Arywn stopped once again, took a deep breath, and looked over the audience before he continued. "Our," he stopped again, realizing his mistake, "*their* remains were sent to the depths below; their spirits remained locked in stone, until now."

Several dozen troops that had been sent to the Hroletariat had begun barking orders at the onlookers. Additional soldiers had cordoned off the doors to the cubicle housing Arwyn and his six ministers. "Come out with your hands in the air!" commanded the sergeant.

In desperation, Arwyn made his final plea. With beads of sweat pouring down his forehead, the great king continued. "Friends, our civilization has come a long way since the Great Cataclysm, but all is not good. Our great kingdom is divided as never before. Material goods are never a worthy trade for freedom, and the false promises made to you by our leaders are simply unrealistic."

There were boos and catcalls from the back, and an

explosive charge blew open the heavy metal door behind which they were standing. But none of it drowned out the deafening sound of thunderous applause and shouts of approval that were heard far beyond the Hroletariat that day.

Finding new strength, the king concluded his speech. "I have come to you with a heavy heart to ask for your help." Three billion years encased in hardened granite had not dampened his spirit nor his resolve. *Could his quest be accomplished without bloodshed or armed rebellion?* This was something he had not anticipated. Time and space were his enemies, but the people, his subjects, were still his ally.

But it was too late! Having gained entrance to the small cubicle, the police, with weapons drawn, moved stealthily up the stairs to where the great king was seated. Leveling their weapons at Arwyn and his small band, the commanding officer gave orders to shackle them and take them into custody. Then something strange happened: their bodies shimmered and trembled, then abruptly changed from solid form to wavering particles of dust, like a mirage. In a matter of seconds, the seven members of the royal family went from material forms to spectral apparitions. The king was the last to fade, his sunken eyes belaying a message of love and understanding. In 45 minutes, he had conveyed his story and accomplished the impossible.

Epilogue

The nation was reborn; conceived in justice, reason, and fairness for all, it once again took center stage.

Three billion years in the making, in a time span inconceivable to humans, the unthinkable had taken place. The ancient rulers had regained their rightful station as the true leaders of this great nation. A new regime committed to the values of the Kitroio-Riehke family once again ruled the lands.

King Arwyn and his six loyal ministers were released from their granite prison and allowed to enter the realms of the Heavens, to sit on the throne of the Lords of the Universe.

# # #

Richard Marschall is a graduate of Towson University in Maryland and Western Illinois University. His works have been published in numerous anthologies, including those from the Space Coast Writers' Guild and the Scribblers of Brevard. His book, *First Call – Poetry for the Ages,* was published in March, 2018. It was followed by *Quarrytown – The Teenage Years* in April, 2018. His first series, *Justice Unbound - Book One: Gehenna,* was published in June 2018.

Juno arrives at Jupiter on July 4, 2016, after a nearly five-year journey. Juno was launched from Cape Canaveral on Aug. 5, 2011, and has a principal mission of understanding the origin and evolution of Jupiter. (NASA)

In the long run men hit only what they aim at.

Henry David Thoreau

Out There

Ashley McGrath

The 1969 moon landing
Gave the United States high standing.
Fifty years later, space still does excite.
We've watched many rockets launch into flight.

Thinking about space makes us feel small.
The night sky never fails to enthrall.
On clear evenings, we gaze at the stars.
People have talked about flying to Mars.

We know more about space than before.
Yet there is still a lot to explore.
Of what are we not aware?
So, let's keep looking out there.

# # #

Ashley McGrath is a call center quality analyst for J. Lodge, a Cognosante company. A lifelong resident of Brevard County, Ashley published her autobiography *UnabASHed by Disability* (available from Amazon.com) in 2014. Her writing has been included in eight other anthologies as well as local publications. She has a star named after her in the International Star Registry. She is a past treasurer of the Space Coast Writers' Guild.

This picture of Neptune was produced from the last whole planet images taken through the green and orange filters on NASA's Voyager 2 narrow angle camera. The images were taken at a range of 4.4 million miles from the planet, 4 days and 20 hours before making its closest approach to Neptune on August 25, 1989. (NASA/JPL)

Now is the time ... for this nation to take a clearly leading role in space achievement, which in many ways may hold the key to our future on Earth.

President John F. Kennedy

ZENITH CITY

J.P. Osterman

Spacecraft *Zenith* launched from Europa's hypothermal pillars at 14:32:02, Earth time, 6/2/5323 Earth date. Monitors on off-Earth colonies displayed twelve thousand humans entering and activating stasis pods for the 13.8 light-year journey to the Ophiuchus constellation, a triple planetary system harboring an Earth-resembling Wolf 106b planet, aka Wolf. While humans have been hibernating in four Level-1 zones for twelve years and six months of the 13-year spacefolding voyage to Wolf, synthetic humans – anthromorphs (morphs) – have been monitoring and repairing *Zenith* (Zenith City), which the omnipotent, omniscient, and omnipresent holographic AI, Delphi, directs via quantum computing.

Fusion engines mix dark matter and neutrino anti-matter to propel *Zenith* to Wolf. Once there, *Zenith* will land on a world thus far blessed as human habitable. Only Delphi can to access the necessary modifications hibernating humanity might be forced to make to adapt on Wolf. After all, nothing is certain until the first select humans revive and can survey the exoplanet. Meanwhile, as *Zenith* crosses the critical *Point of No Return* to Wolf, and as Delphi emits the first code to open the quarantined file, a quantum peel rings through *Zenith* at nano-zephyr speed, awakening the morphs.

***

On Zone 2, in which 4,000 humans hibernate, AI

3227 gasped and sat up straight out of quiet mode. She checked her translucent wall grid that functions as her alert monitor, quantum-computing hub, time informer, and transport beacon. "I – I – think you're giving me a wrong wake-up call, Delphi." Walking to the center of her 8x8 quarters, she waited for Delphi's response, shuddering a bit. "I'm off, *really* off, Delphi. Where are you?" She rubbed her forehead. She believed she felt dampness and glanced at the dark ceiling, rippling an effervescent, milky energy mirage as quantum-computing processes and exchanges passed throughout *Zenith*.

Hologram Delphi appeared as a mirror image of 3227 at the center of her quarters. Delphi has shoulder-length brown hair, is wearing a shiny blue jumpsuit, and has 3227's vitals illuminating bright-yellow on every synthetic body organ.

"Everything about me looks normal," 3227 read, "but, why the early activation, Delphi? I had five more minutes of quiet mode coming. That means tonight I'll have *fifteen* minutes instead of *five* minutes of downtime before retiring." Trying to calculate more consequences, she paced her small room, empty except for her elevated body pod and illuminating wall-grid monitor. "Fifteen minutes of down time that's wasted time away from my work station, Delphi." A thought unlike any other before occurred to her. "Am *I* the only one? Did this early wake up happen to just *me*?"

Muffled noises resounded outside her steel sliding door. "3227, an L-train is arriving on platform 1 to take you to Zone 2's charging station."

"First, Delphi, answer me," 3227 demanded, her hands on her hips.

"You have three minutes and ten seconds to select your outfit, 3227," Delphi said.

3227 walked up to Delphi's hologram, its face void of expression, its eyes appearing as lifeless blue marbles lacking acknowledgments. She had programmed her Delphi hologram as an actress from Earth, Cheryl Tiegs. In their free time, morphs participate in virtual shows and learn about history to help humans on Wolf. "I demand an answer, Delphi, now!" She walked around her. She thrust her hand into Delphi's unreceptive hologram that didn't waver, or flicker, or disappear, but began meeting 3227 face-to-face in response. "You have two minutes, forty-five seconds, 3227."

Emotion filled 3227, and she kicked her body pod. "Who *else* did you wake up early? I want to know, *now*. Answer me!"

Delphi's hologram flinched. "All zones are affected."

On 3227's wall grid, *Zenith*'s spacefold trajectory illuminated. A giant red band lit up, close to Wolf, enclosed with the words, *Point of No Return*. "*Zenith* has passed the Point of No Return."

Another thought triggered in her as an automatic response to gulping down air. "Then, I'm not the only one to activate early. The *entire* craft must also be off by five minutes." She shivered, and a bead of water slipped down her forehead and dropped into her eye, blurring her vision of Delphi in front of her.

She heard a scream outside.

She felt a sting in her core. "I'm, I'm not myself, Delphi."

A new reading radiated from the mirror-image vitals. A yellow string appeared at the base of the skull point.

"What's *that*?" 3227 asked, stepping back, the sting radiating into her chest. She could feel something there now, a cold sensation she had never experienced. "Something's wrong!"

More screams resounded outside, with a squealing noise of the electric train stopping on the platform in the distance. 3227 had five minutes, with a time stamp blaring on her wall-grid: *07:35*. The L-train would depart in ten minutes – one of five electric-driven trains stacked in crisscross levels to transport all morphs in Zones 1 through 4 to their work stations. Her work schedule on Level 4 Center was always the same: 09:00 to 20:00, with an hour of charging time at the center's round pod station from 08:00-08:45 and 20:15-21:00. Then home to her quarters for a night of quiet mode.

"*Zenith* surpassed the Point of No Return," her Cheryl Tiegs hologram returned. "3227, you have not changed your clothes. 3227, change your clothes."

3227 looked down and then stared at her reflection in the wall grid. She had on a bodysuit, but that was not her uniform for monitoring life support stats throughout Level 4 of Zenith City. "Set me to Uniform Red, Delphi," she ordered, and her sleepsuit altered into a long-sleeved, red dress. She touched her peach-colored cheeks, everything outside and inside synthesized on Earth. She ran her fingers through her silky brown hair and wondered: *Who* manufactured *this*? Out of what? Her programming reminded her about the hibernating humans and their short life spans. Every morph was assigned to a corresponding human as a helping tool. More words came into her mind: slave and servant. She clenched her teeth. *She* would be *someone's* slave. A metallic taste welled in her mouth. Where *is* that person in right now? A man? A woman? I want to see that person, right now! With tight fists, she realized the truth. She could not know the human to whom she would be assigned until sometime after disembarking and preparing a habitat for him or her on Wolf. Only Delphi knows his or her whereabouts, how he or she looks, and who they were on Earth.

Her shoulders pinched in a quick ache, and she felt a click in its artificial sinew. She had two minutes to catch her train, time that began to feel so real and fleeting in the smooth contours of her face – contours that might age one day, or maybe remain the same until someone would decide to deactivate her. Then, that will be all for me, she thought, the end, nothing, no more.

A feeling of self-awareness erupted as a powerful memory of a lava flow she had seen while studying the deadly climate kill-off on Earth from 2904 to 3200. She turned around, examining her body, her physical appearance inciting an internal riot of emotions. Am I pretty? Old? Leaning wild-eyed into her image, she touched the corner of her eye from where that speck of water had evaporated. She looked up. There was no condensation on the ceiling. She glanced around. No faulty *Zenith* water relay or vapor pulsing through the craft to instigate repairs or for recycling. An emotion like the distance between her and the dark space outside Zenith flooded her. "I think, if I were human, I'd cry right now, Delphi."

"3227, you cannot miss the train." Delphi approached her.

She retreated. The sound of the number made her taste that metallic liquid again in her mouth. She wished it could be blood. "Stop calling me that!" She struck her body pod with her fists. "Stop it!" She needed something – for the first time, a driving force. "I'm sick of you calling me 3277!"

Delphi went into *standing-straight* mode, attempting to run a streaming code into 3277, redirecting her. "3227 – "

"Stop calling me 3227!" She raised the pod, toppling the pillow. "I'm *not* 3227!"

Delphi immediately discerned the unprecedented action. "3227, I am attempting to make a shift in your consciousness."

Breathless, 3227 called, "Open!" Her sliding door swooshed open to reveal a moving walkway, bustling with hundreds of morphs on the way to their L-trains. "I have a name, hear me? And it's *not* 3227!" Quickly, she scurried between other morphs, losing herself in the crowd. Peeking ahead, she noticed a few morphs ducking up-and-down, waving at her. They also appeared frightened but were trying to get her attention without alarming the Delphi system.

She checked her rectangular skin processing pad embedded in her forearm to see if she could ID them. Nothing. She had coworkers with whom she socialized, but no acquaintances on her zone. "How do they know *me*? I don't know *them*!" She maneuvered closer to them, closing in on the embarkation platform. Then she remembered the screams she had heard from her quarters, and she began searching for the chaos.

Now that I'm experiencing new sensations and feelings, others around me *must* be experiencing the same things, too, she thought. She darted into the L-train, ducking low among other morphs through which she knew Delphi would be searching for her. She had only seconds before detection. She recalled metal conductivity, streaming, and blind spots. My only chance is to find a corner and huddle, she concluded. Every strategy felt indelible in her brain. Remain low, get to a corner, and alter into *nude* mode. That would mean, no clothing. But colors act as detectors, and if I can make it to Level-4 Center, then I can take a lift to Level-1. I want *my* name, and I want *my* life. Only the humans can grant me that. I'm going to do what no one's done before. Meet them personally to get what's due me, or find the others and have them help me overthrow Delphi so it will take me to the humans. Sidling through a few bodies, a morph approached her, towering inches over her head.

"I'm 3228," he said. A black man, he had a gruff voice, steely blue eyes, and mounds for biceps.

"You know me?" 3227 stood on her tiptoes to peer

past walls of morphs, some appearing to shield her. A thought seeped into her awareness – a feeling of kinship and connectivity as more morphs approached her, some surrounding her in protection. "What is going on?" she whispered to 3228.

He had a blank stare as he touched her arm. Morphs never touch one another! She jerked away. "I want my name. I woke up this morning with your number repeating in my mind."

"So did I," a female morph with brown skin said. "I am 3230."

"I woke up with that thought as well," another female morph with short white hair added. "I'm 3225. I was drawn to you, 3227. *You* know where we're all supposed to go for our names."

"Me? All twenty-seven of you, drawn to me? I don't understand!" She fell against the corner – the outside tunnel lights flashing white and yellow on the faces of so many desperate morphs in search of two simple questions: What's *my* name, and how can I have *my* life? She had another realization. Some change must have activated on Zenith. Some human must have created it to initialize at a specific point, changing some morphs, and making *me* like a magnet for them.

"Delphi won't help us," 3228 exclaimed, his head lowering. "I had *that* revelation when I tried to ask it for help, but it ordered me to report to the turret for examination." He heaved in several breaths of anxiety and despair, emotions so raw for them all, and uncontrollable. "No morph's ever even been to the

Examination Zone, but we know what it is, a reprogramming lab!" His saucer-blue eyes appeared about to burst. "Uh, no way will I let that Delphi down there whittle away at *my* mind! So I ran, ran as hard as I could, right into, well, into *you*, 3227, like I was meant to be with you." He shuffled in his feet, pointing at some blank and yearning faces around them. "I don't know what's going on with me and the rest of us – "

"Me either," 3230 interrupted. "But I'm with *him*. And since he's with *you*, 3227, *I'm* with you."

"Count me in!" another called.

"Me!"

"And me!"

"Don't forget me back here!"

Echoes of "me" peeled in shrieks and cries through the car, from the mouths of all twenty-seven anguished faces, until she felt tall, powerful, and energized, imbibing their strength and determination. She grabbed 3228's arm. "Are you sure, *absolutely* sure, that Delphi *won't* grant our requests?"

He nodded. "She won't. She'll send us all to the turret level, and then we'll be no more." He made a motion of slitting his throat. "When this train stops, we will *all* be stopped by other morphs who will lead us to the turret." He leaned in for a hard terrified whisper. "I *saw* the future!"

In a long pause, as the train shifted, and outside lights raced inside as yellow and white stripes on their heads, all twenty-seven morphs began murmuring in confusion and fright. 3227 put herself in their shoes. All of their emotions were new, raw, intense, and driving them. "3228 is right," a morph called out from the back. "Delphi won't give us names or set us free from our assigned humans. But I'm telling you, I am 5444. I have a voice, and somewhere, a name, and I have a *right* to my *own* life."

3228 stepped closer to 3227. "We are trapped, and our destinies predetermined. Do we want to live, possibly for eternity, as *only* numbers and slaves?" He pounded a dent into the wall, a light fixture popping and crackling. Other morphs that had not yet experienced their calling began staring at them as if summoning Delphi.

While listening to her comrades, 3227 nodded at the expressionless morphs, the gesture conveying peace and calm.

"Delphi can name us, and free us!" several morphs chanted. "We want names!"

"And freedom!" more called.

"Let us proceed to Delphi," another said softly, "or the humans."

The entire crowd affirmed 3228's conclusion. They were only numbers, and they would remain numbers and *only* serve humanity.

"Over my *dead* morph skin!" 3227 shouted. She felt wronged, slighted, cheated, trapped, and bitter. Great injustice had been thrust upon them before they had exited their molds. Clenching her fists, she raised her hands. 3228 lifted her into the air like a beacon. "When this train stops in five minutes, we *will* fight against Delphi and *any* morph who tries to stop us."

"I want my life," one of the twenty-seven morphs shouted.

Cheers and affirmations resounded. Meanwhile, the other passengers stood motionless on the opposite side of the train, their faces and eyes staring into space, in deep connection with Delphi, or *Zenith*.

Morph 3119 slapped an opaque patch over 3227's wrist processor. "I am 3119," he said in an Asian accent. "I am in charge of wall maintenance, and this fabric can prevent Delphi from accessing us." He ran to a wall and began slicing off the fabric. "Do what I do!"

Morphs began frantically peeling off pieces of the wall and covering their wrists.

A few times, Delphi appeared to interrupt them. Then, the holograms ceased.

3228 ripped out three steps, created a platform, and took 3227's hand as a king would guide his queen to her throne.

"If need be, there *will* be a war!" 3227 yelled, and her comrades applauded, calling her their leader.

***

Meanwhile, at Central Zone 3, on the western side of Zenith City – over manufacturing equipment, a Noah's Ark, and a dehydrated food supply – several morphs halted work at their terminals to stand at attention. Delphi had initiated an intensive processing stream with them. Zenith City had no need for police, or the law, as on Earth; however, given the sudden rebellion, Delphi began a fastidious implementation of a law enforcement paradigm: a police force would manifest with select morphs on Zenith City. Delphi picked the chosen.

As Zone 2 quarters morphs 3,000 through 5,999, Zone 1 houses morphs 0001 through 2999, Zone 3 houses morphs 6,000 through 8999, and Zone 4 quarters morphs 9000 through 12000. *Every morph for every human* is the motto above the arched airlock leading to the 1,550 shuttles that will one day launch for Wolf.

Delphi's voice resounded the time, "Seven forty-five a.m.," through the domed work area, droning with transparent monitors streaming Zenith City's vitals, *Zenith*'s hull integrity, and the space craft's interior life-support readings.

Morph 6441 experienced a revelation. Peering into her numerical-streaming transparent monitor that had unleashed a hypnotic interaction, 6441 looked at her reflection from head-to-toe. No scans, she thought, but there should be functioning vitals streaming alongside me. "Delphi, where is my holographic double? I need my vitals, or sleep mode will activate, and I will be useless." A quick sharp jab fluttered in her core. "What was that?!" She tilted forward. Her vision blurred. Her hands trembled. Her breathing quickened.

Delphi appeared as a brightly haloed male doctor dressed in white with a stethoscope around his neck. "You are experiencing fear, 6441."

She breathed deeply. Nothing changed. "This fear is *not* dissipating, Delphi." She struck her hips and torso. "Get it off me! I do not like it!"

Delphi walked through a maze of twenty-seven morphs that had stopped their work and went into apparent sleep mode. She touched them one-by-one, and they reactivated, inspecting their arms and legs as if they were newborns testing their environment. "Breath, 6441," Delphi said in a low tone, "and fear will dissipate."

She glanced at her blond muscular reflection in several transparent monitors around her. She couldn't believe her eyes. "This is *me*? What I look like?" Pinching her cheeks, she nearly kissed the information-streaming screen. "I did not notice these things – "

"Traits," Delphi interrupted as 6441's fellow morphs followed her example and began examining their faces. "You are noticing your traits. You are becoming aware of how you look and feel."

"Why?" 6441 had been in constant instruction mode with Delphi, but these moments felt new and unusual. Several scenic views from Earth's history sieved through her mind, and a feeling of tranquility moved through her. Delphi had initiated a change, and she and the others were experiencing a calling. "I think I understand," she said, walking among her comrades. "*We* are students, and *you* are the teacher," she said to Delphi, and a tender emotion quenched her fear until a sudden emptiness usurped the panic attack. "What are we supposed to do now though, now that you've infused us with this calling?" She began experiencing various emotions, and Delphi streamed a neurological explanation into her consciousness. Glancing into the eyes of her companions, she realized they were experiencing the calling as well. "Delphi, now what?" She folded her arms. Frustration twisted into impatience. *Will I have to wait for it to tell me my every next move? Oh-oh, I bet she heard that!*

Delphi answered her, pointing for her to move quickly toward the L-train platform to embark on the next train in two minutes.

That meant Delphi had *not* discerned her thoughts. What did all this mean? She was changing. How? As she and her twenty-seven comrades walked quickly to the L-train station, 6441 accessed Earth's history and researched her new attributes. *Emotions, sensory experiences, the ability to question one's environment, the capacity for empathy, and being mindful of one's uniqueness among others.* Hmm, maybe this means I'm getting closer to being a human, but without being physically human.

Within seconds, she and her companions boarded Zone 3's train 3.

"Where are we going?" Asian morph 8973 next to her asked. Several more morphs asked her the same thing.

"What are going to do when we arrive at our destination, Delphi?" others asked.

Delphi illuminated wall monitors. "Twenty-eight morphs are on their way to deactivate me. That is all the information I have."

A pathway illuminated above them. L-Train 3 was on a trajectory to Zone 2's Central Platform, one of four roundabouts on the ship. After docking, train 3 could rotate and move vertically two ways: four levels to the top, to the lookout dome and Delphi's processing center, or down four levels to one of the four quadrants holding four thousand humans in stasis.

"Deactivate *you*?" 6441 believed she saw white stars. "But they cannot do that. No one can deactivate you, Delphi."

"Yes!" a morph in the corner agreed. "You *operate* Zenith City. Deactivating you would kill us."

"Stopping Zenith City is against *any* morph's constitution," 6441 said, still not comprehending how any morph could become violent and separate from the community network, killing everyone and the mission to Wolf.

"It is *impossible* for *any* of us to interfere with Delphi in *any* way," another morph shouted.

6441 experienced a revelation. She rubbed her hands together so tightly that she believed her fingers might rip off. "Wait!" She faced them all. "A morph *can* interfere with Delphi."

"How?" several questioned.

"By becoming, *somehow*, and at *this* time, aware, as all twenty-eight of us are *right* now, of pain," 6441 replied, still rubbing her hands. The agitation wouldn't go away. She was experiencing sensations, several at once, for the first time. "This is so painful!" She blew on her fingers, circulating the vital red fluids.

"Look at Train 1's path," a morph in the corner pointed.

High above them, as outside tunnel lights beat to the time of muffling clacks on their L-train treads, a trajectory flared yellow and red for danger.

6441 gasped. "Looks like our trains will collide! Is that path right, Delphi?"

Delphi appeared in several holograms throughout their train, assuming famous historical figures from Earth's history that various morphs had preprogrammed for it to manifest: Abraham Lincoln, Gandhi, Mother Theresa, and the last global leader on Earth before their launch from Europa, President William Pope. Each hologram began telling its host, "Calm down," "Don't fear," and, "You can defeat the morphs on Train 2."

"Defeat them!" 6290 exclaimed.

"What does *defeat* mean?" 6110 asked.

More panic ensued until Delphi appeared as one giant hologram at the center of the train. "6441 will be the first to disembark. The rest of you will follow her."

"Me!" 6441 fell back into 6290's shaking arms. "I don't know what to do?"

"You can, 6441. I will give you instruction when the two trains meet," Delphi said, her shining god-like face flaring an artificial peach.

6441 felt her feet turn numb cold and her toes sting.

Something didn't ring right in her ears, twisting as if the numbers in her identity smashed, disappearing. "Why me, Delphi?"

"6441, prepare to receive instructions upon disembarkation," Delphi ordered through intense yellow eyes.

"Why not call me sixty-four, forty-one, Delphi?" she asked squinting, turning away. She didn't want to gaze into those blazing eyes and not receive their instruction. Delphi *could* kill her.

"6441, receive your instructions for a law enforcement code," Delphi commanded, louder than before. The morphs approached 6441 in a protective instinct.

"Why not call me sixty-four, forty-one, Delphi?" she said at the top of her lungs.

"6441 – "

"Stop it! Stop with that damn number because it's a number, and I'm *not* – "

Their train skidded to a sudden screeching halt, interrupting her.

Two sets of sliding doors swooshed open. 6441 stepped out slowly, tiptoed alongside the train, and waved for her fellow morphs to duck down and follow her lead. Through the steam, and across the two sets of tracks, she searched for the rebel group. She could see darkness all over the steely cold platform floor, the outline of the rebel's west-bound train, and a dull light emanating from the interior of their train. "They should be near here and hiding, within several yards, because the train appears empty."

Delphi continued walking alongside 6441, whispering directions to her. Its eyes were bright blue scan lasers.

Trying to steady her breathing, 6441 touched the vibrating train that automatically descended from gravity thrusters upon her touch. "The train *appears* empty," she whispered to the others. "What's going on here, Delphi, because you should know if they're inside this box or not."

6441 grimaced at Delphi, a hologram her angry fist would ease right through. Still, she realized the facts. The moment Delphi's blue-flaring eyes fixed in one position, it would have a lock on the rebels, if they were anywhere around the train. 6441 quickly imagined what to do: run to them, grab the first morph her arms could embrace, and hold on tightly, apprehending him or her.

"Come out! Where are you?" she called, her throat dry from the air smelling of salt and coolant. Delphi had been filtering the platforms, but it takes time for acrid tunnel air to dissipate. She wiped her stinging eyes, signaled for ten of her companions to join her, and slumped behind a platform wall dividing the exit from the entrance. "When I call *go*, we charge the train, tackle them, and capture them."

"I do not hear *anything* though," 6290 whispered. "Are you *sure* Delphi led us to the correct platform?" He appeared spooked by the eerie sounds, new smells, and burgeoning sensations.

7110 peered up with his piercing green eyes, then ducked away in fright. "I don't see *any* morphs inside either. But a few morphs behind me said they have something *really* important they need to show you *right now*." He gestured through a distant dark location with ghostly shadows wavering in dull light.

"That's the emergency energy stations. Maybe a few of them are drained and are recharging." 6441 felt a surge of victory, yet worried about the amount of power she had left. Checking her wrist band, she saw her green safety light and calmed down. She passed the word for her companions to manage their storage supplies. "We can't lose even one morph," she said, flexing her biceps, believing herself impenetrable. Seeing a thin panel on the side of the train in a state of automatic repair, she asked 7110 and 6290 to help her peel it off. "A perfect shield for us," she said, rounding the putty edges that cooled and solidified.

"Yep, that's the place a few morphs behind me said they thought they spotted bodies," 7110 said. A tunnel wind wafted around them, raising their hair in a static cling.

6441 squinted hard at the dismal dark location that felt like a long, drawn-out loneliness: a few shadows wavering in a dull yellow glowing distance. A red square appeared high above the phantom figures. Delphi illuminated architectural plans in front of them.

"It's an exit into an evacuation tunnel." 6441 motioned for several to rally around her and hold up the shield. Delphi activated the shield's transparency capabilities so they could see through it and discern every object around them. "The rebels could have fled. But still, a few of them *must* be around. They have to realize that Delphi would send someone to stop them."

6441's insides sickened – if she had any insides, other than organic nanotechnology making automatic repairs and processing all her vitals. "I'm going in." 7110 pulled her back. "Nope, let me go. Someone has to take the first leap." Her chest revved through a strong pulse, and she handed two morphs the shield. "Watch for my signal!"

Crawling around the dividing wall separating the embarkation platform from the dark interior, she skidded on the terminal floor, landed on her stomach, closed in eyes in terror, and felt the skin on her cheek peel in heat and water. Looking up, she spotted three bright lights and listened for breathing and movements. Nothing. She pressed her hands to the cold damp concrete and stood up, tall and sure. "No one." She waved for the others to come out of hiding. "No one's here, Delphi! What the hell?"

"Yes, there are morphs here," said 7110, and he illuminated several lights around the emergency energy stations – body-sized booths with body contouring sockets. Four morphs were in various states of reenergizing. One morph had not completely separated from his charging booth – his back glowing in a quantum-entanglement with several hardware sockets.

6441 and the rest of her comrades joined him in astonishment. "They're *all* in stasis!"

"Unleash him, Delphi!" several morphs shouted.

6441 knew the outcome if they couldn't free him: a bending overload and internal fry. She had to act fast. "Hand me *anything* rubber." She glanced around the terminal. Another morph activated two breakers, illuminating the domed terminal.

"Over here!" a morph called, ripping at the train's bumper, its nanofabrics sparking in retaliation. Heaving and pulling, several morphs tore it off and handed it to 6441.

Running to the limp morph, she and 7110 thrust the rubber appendage into the yellow-hot quantum stream, and the morph fell into 6290's arms. As several morphs inspected the morphs inside the energy stations, 6441 examined the morph. "He's like one of those old-Earth mannequins."

Standing, she approached Delphi. "How do you explain this? A faulty separation is *never* supposed to happen! It's one-hundred percent failsafe for all of us." Despondence, like the slow-flowing sappy nanoliquids repairing a floor crack, flowed through her – a possible ending that could happen to *any* of them the next time they had to energize. "We're in trouble," she said, covering her lips. "What do we do now? It's obvious Delphi is off somewhere else and making serious mistakes."

Delphi disappeared and reappeared alongside the lifeless morph, and in front of the three other morphs still connected to the energy booths. The booths had halted energy infusions, and the morphs had lifeless faces and limp bodies. Delphi's persona was flat and remained nonresponsive.

Fending off a headache, 6441 touched her forehead. "This is so impossible!" Anger vented through her hands, she struck Delphi, but her fingers passed right through the hologram. "Well, *that* was useless." She kicked it. Only light fell on her foot in response. "Delphi, get with us here! You want us to find the rebels, but you're not giving us hope that we'll even make it to them."

Having searched the area and discovered no other morphs, 6441's companions joined her. "Delphi!"

Overhead in the distance, at the main entryway and above several sliding doors, she noticed Zenith's trajectory. "We deviated off course." She gestured for the others to confirm her discovery.

"We sure did!" one said.

"Slightly," several agreed.

"Two kilometers, but then we assumed the correct course," 6290 said. "I'm a monitor at Level 2, the Navigation Zone.

"When did the deviation happen?" 6441 asked.

"Ten minutes and thirty-three seconds ago," 6290 replied, showing 6441 her special icon to access the zone.

6441 took a deep sigh and shook her head in a helpless gesture. "Oh."

Every morph stopped, instinctively knowing she'd

have the answer.

Her shoulders sulked. "That's why Delphi's mostly been useless to us." She walked toward the train on which they believed harbored the rebels. "Some program from Earth must have activated at the Point of No Return. And as we've been passing other specific points – "

"That no one is aware of," interrupted 6290.

"Yes, and not even our all-knowing AI here..." She thrust her hand through Delphi's unwavering hologram. "Everything has been changing around Zenith City."

"Including all the morphs," 7110 said, touching a morph fixed to one of the energy pods. He had disengaged a large wall panel, severed two circuits, and began releasing the other three morphs from their energy pods. Freed, they remained fixed, stationary and rigid.

"They're inactive," a morph called out from the dark. Several voiced their concern that they, too, might deactivate at any time. Panic ensued. "Is that all we are to this damn place? A bunch of expendable solids and liquid pulp that the humans can one day use and throw away?" Morphs were growing angry at the sight of their comrades.

"Is *every* morph like this?" several shouted.

"The humans did this to us!" a morph cried.

"No, it's a horrible system virus!" another screamed.

6441 raised her hands and faced the morphs. "Shut up! Calm down!" In the quiet, she breathed long and hard. "If anything terrible was going to happen to us," she began as she walked to the train, "it would have happened already." They sighed in agreement. "But obviously, we *can* conclude that *all* morphs on *all* levels are in the same state as these here." She gestured at the inactive morphs that 7110 had lain gently on the floor. "For some reason, Delphi, or the *Zenith*, initiated a program of *complete* change. It affected us differently than them," she touched her chest. "Them *and* Delphi." She paced the floor. "Makes sense. We veered off track to Wolf a little, and then morphs freeze, and then we and the rebels change."

"Or vice versa," said 6290.

"Maybe Zenith City has *completely* taken over the ship, like it's coming alive or something," 6441 speculated.

"The rebels," a morph called. "Maybe *they* know what's happening. That's why they rebelled."

"Or they're on their way to find answers or fix these problems," another morph said.

"They *could* be good, while all along Delphi here has been telling us that they are bad," cried another morph.

6441 thought her mind might overload in the pressure. "Some *serious* problem, or disaster, definitely ignited them into action, just as Delphi changed *us* to apprehend them."

In the new terminal light, she checked the train insignia on the sizzling, trembling electric track – its bumper half in auto-repair and glowing. After ordering two morphs to pick up the shield, she noticed the train's number and grew angry. "Train 2? *This* is train 2! The rebels were *not* on Train 1." Through gasps and cries of shock, she ran to Delphi. "What the hell!" She wished she could shake her.

"Delphi's made another *big* mistake," 7110 hollered, joining her. His fingers appeared scorched but were quickly repairing.

"You think so?!" another morph added, her raging face penetrating Delphi's hologram. "Our AI has definitely gone bonkers!"

6441 glared at the AI towering above her, the hologram suddenly wavering – a processing deficiency. "What train did the rebels take? How are we supposed to stop them if we can't *find* them?"

Every morph began to ask: "Where are they?" "Where can they be?" "How much time do we have before they reach Delphi's core to kill it?"

"My gosh!" 6441 cried. "I want to know what the *hell* started this whole thing."

"Yeah, Delphi, good question," several morphs said, approaching 6441 closely as if fearing the great AI.

6441 paced the platform while rubbing her hands together. This makes no sense, she kept thinking, no sense at all. Then she glanced up at their trajectory in space – always displayed at strategic locations throughout Zenith City. "Delphi," she said, pointing at the display, "we've passed the Point of No Return, but you have markers on display after that point." She recalled several processing issues Delphi had already exhibited since it had summoned her inside her quarters.

6441 traced a bead of condensation etching a twisting path down a support beam. "As we're passing critical points, Delphi is changing." She bent down, an empty feeling engulfing her. "We've been changing. I think this entire spaceship's changing. How, and into what, I don't know."

"Huh?" several morphs gasped.

"*Every* morph in Zenith City?" 7110 hollered from the back.

6441 walked straight through Delphi's hologram, and it didn't flinch, or disappear-to-reappear, or say anything enlightening. "Well, we *here* certainly have changed." Through sighs and words of alarm, she added, "But I don't know about the rest of our kind."

"Don't forget the rebels," 6290 said, gesturing hard at the empty train. "They ghosted us, and outwitted us, because they had to know someone would try and stop them."

After questioning Delphi, who responded with more questions, 6441 rubbed her stinging sore neck. Before, automatic repairs would have rushed to any of her tender or malfunctioning organs, relieving any discomfort in her artificial muscles. This time, she believed Delphi had abandoned her to an uncertain and unpredictable existence. No stasis for the morphs! Perhaps an early death. But what a wonderful life the humans will have on Wolf! Maybe the morphs were only voyaging on Zenith to ensure that the humans would arrive at Wolf safe and sound. Perhaps we are dispensable! And this is it for us ... the end.

"The humans!" she said, angrily striking the entry pad to Train 2. The sliding doors hissed open.

"What about them humans?" 8888 asked in her Southern drawl, glancing around the dark terminal, her expression one of shock.

"It's so obvious!" 6441 struck the side of her head, a gesture of stupidity. "Why didn't I think of this before?" Delphi had already told them that the rebels had disconnected from it by covering up their arms processors with special nanotechnology. "I know you can't track them, Delphi, but since *you're* still appearing to us, we know they haven't arrived at your core and interfered with you, at least not yet."

"True," Delphi agreed. "6441, I am working to discern the location of the twenty-eight rebels."

***

The platform resonated with L-train steam emissions and resounding departure and arrival stats. The air was filled with the smells of artificial flowers and ocean air, Delphi's practice for when the humans would awaken one day and live on Zenith City until they could build a solid foundation on Wolf.

Delphi calling her 6441 again made her blood *feel* boiling. *I hate that number*, she fumed, *hate it with a passion*! She put herself in their place – what would make morphs rebel, and how might *they* be feeling?

"Wait!" she stopped.

"What?" several morphs asked and then stilled.

In deep thought, 6441 paced the platform. "If I *want* something, or someone, that only Delphi or the humans can give me, I'd go to Delphi – "

"But Delphi is *here,* and functioning," 7110 said in a certain voice.

"If you call what we have here functioning," scoffed 8888.

6441 recalled that only human beings could rebel. It's in their nature to fight one another as good versus evil, if indeed the morphs they are chasing *are* evil. They weren't sure yet. But she knew she couldn't kill another morph – let alone an innocent human. It wasn't any of *their* faults that scientists chose the twelve-thousand hibernating humans to colonize a Wolf planet. It wasn't *their* fault that they were born human and the morphs fabricated.

6441 looked into the questioning eyes of her vulnerable comrades. They were all helpless and in need of her leadership and ability to create order out of chaos. Wherever I've been getting these skills that I've been using for the past forty-five minutes, I have *no* idea, she thought. But I have them, and I seem to know exactly what to do at the right moments. One thing's for sure: I can't kill. And I won't order a kill. Morphs, like humans, have an unfailing moral compass, and not one of them can kill. "Only humans can rebel against a closed system," she whispered, thinking hard about human history. Gasping in a sudden revelation, she put her hands on her chest. "Oh, no!"

"What?!" morphs asked, steadying her on the final step into Train 2.

"The rebels *must* be going down four levels to reach the humans! Or they *could* be there right now!" 6441 checked the time.

"They probably took another way down from this level not too long ago," a morph gestured at Delphi's station clock, displaying all the departing and arriving trains.

"No way could they break through *two* vaulted security systems to reach them." 6441 motioned for them to embark Train 2. The doors swooshed shut after the last morph jumped inside. "Descend, Delphi," she commanded, setting in motion the train's anti-gravitational thrusters. Delphi lifted and turned the train 180°. They began a rapid descent to Level 1's human zone, housing 4,000 stasis pods. Glancing at her wrist processor, 6441 saw the time: 09:29. She knew their intelligence. Like her comrades, the rebels were all workers and had intensive training in security hardware and software. They would also be smart, *very* smart. "Most likely, we'll catch them working to breach the second door." She had the shield and began picking off small wads of nano-putty.

7110 peeled off material from the overhead lights and ripped a piece of copper plating off the baseboard. The train tilted and then steadied. "Little bombs," he said.

"You got it!" 6441's fingers stung, and she sucked

viscous blood, or what humans called *life pulp*. "Throw them on the floor or at their feet, and we create a distraction."

"Sounds disabling," said 6290. "How many do we get?"

"Enough," said 6441, breathing pungent air, stinging her nose. "As many as we can roll up in two minutes.

The train surpassed Level 3.

"Level 1 in three minutes," Delphi announced, disappearing and reappearing among the morphs.

Again, the train tilted ten degrees, pitching them to the left.

6441 grabbed a pole and caught two putty balls that nearly exploded in front of morph 8888's face. "*Whew*, that was close!" She calmed her down and noticed Delphi's flickering holograms.

"What happened?" several morphs asked, regaining their balance.

"*Something* happened outside!" a few morphs cried.

6441 sat back, wondering if she could handle another surprise, yet the *Zenith* was hitting them with a *world* of surprises, and they were being forced to adapt. "Well, the train's online, and we haven't exploded, so *Zenith* must have passed another critical point."

"6441," Delphi began. "Please watch the explosives in your hands."

"Oh shut up, you useless thing!" She fake-threw a putty ball, imagining it striking Delphi's hologram like a miniature firecracker. "Someday, I hope you turn into a solid, then I'll really let you have it," she scoffed.

"Feeling *mighty* agitated, huh, 6441?" said 7110. His green eyes had a gentle tenderness.

His new accent astonished her, as all the morph's nuances and personality quirks were surprising and puzzling her. "No," she flinched, "I don't know what I hate more right now, that awful number the humans assigned me, or the useless AI calling me that."

Looking over all their heads, she saw their holographic numbers – identity marks. "Numbers," she mumbled, "numbers *suck*! Who the hell would have thought we'd all be just numbers."

***

Approaching the station, Train 2 slowed and jerked three times. 6441 slid putty balls into an emergency backpack she had emptied of supplies. Then she recalled battles on Earth, or maybe it was Delphi infusing her with an upstream of knowledge? Whatever the case, she remembered who won, who lost, and the best battle strategies to implement right now.

"Ok, when we disembark, I want you to form seven lines," she said, feeling strong and in command. "That's seven lines for the twenty-eight of us," she gestured. "I'll be at the center in the first line, or as I'll call it, a wall. The rest of you figure out who will lead the other six walls behind me." They began dividing. "That's right," she coached. She brushed back her bangs, feeling the sweat that had never before appeared. Have I been hit with something, she wondered, glancing at the ceiling.

She felt her hair. Dry. Smelling her fingers, she then noticed lines. Prints? Wrinkles on my knuckles? *What the –*

The train stopped. The doors opened. She wiped her sweaty forehead on her sleeve. "Ok," she said through a rush of, what must feel like, human adrenaline. She focused on the outside platform.

"See anyone?" 7110 asked, breathing rapidly over her head.

"Just the dome lights," she gestured upward.

"It's so quiet," several said.

"I know," 6441 returned, "so they *must* be at the medical facility." She motioned for them to line up and opened her backpack, exposing the putty bombs. The leaders behind her copied her. "Let's go! Delphi, you lead."

Delphi illuminated the zigzag path to the medical facility.

"Can it get a lock on their position?" 6290 asked.

6441 believed her arms and legs might fall off from the numbness of squatting and standing as they maneuvered through the illuminated hallways. "Nothing," she said, continuing to wave them forward. She asked for the shield, and they put it in her palms. "Hold it." She gave it to three determined morphs on her right. "You run up alongside me when I signal." They nodded.

Then Delphi stopped.

6441 almost penetrated Delphi's hologram. "Noises!" She motioned for everyone to stop. Ahead, she could see a flickering light, hear bustling, and make out intermittent sawing. Step-by-step, she tiptoed closer to the voices, their words becoming comprehensible.

"Pass the word!" She knelt, two explosives in her fingertips. Delphi was displaying a fifty-yard distance to the last sealed entrance to the humans. Alongside them, she smelled an acrid scent. Lighting the walls, she pointed at two smoldering sliding doors, their wall panels also charred and sparking.

Several morphs behind her stated their position: the rebels had breached the first entryway to the humans some time ago. From the sounds ahead, she realized they didn't have much time to stop them before they could kill humans, *if* that was their goal.

"The rebels *must* be almost inside!" Altering the soles of her shoes to *rubber* mode, she had those behind her do the same. Walking faster through the long illuminated corridor, she sensed her racing heart, her hot expanding lungs, and her muscles tighten into fighting pistons.

"Charge!" Running, she stopped and gasped when she saw a group of morphs in defensive mode. In the distance, the entryway doors appeared almost separated. Whoever had been working to break inside quickly stopped. She held her fist in the air. "We have explosives, and we'll use them if we have to." Her comrades holding the shield came alongside her. "We *will* fire! Surrender, now! Stop what you're doing!"

The group lined up yards in front of her appeared worn and disheveled. Had they used their energy and needed more? She lowered her explosives and quickly checked her power status: *6 hours*. Most likely, some of her comrades were running low by now, and soon, they'd *all* be in need of re-energizing. The harder the work, she realized, the more the power drain. And they were not used to exerting themselves, which they had been doing the past hour. Slowly, the line in front of her parted.

A thin woman with short brown hair approached her group. By the expressions on her comrades' faces and their body language, 6441 discerned the woman to be their leader. "I'm 3227." Her blue and white uniformed was smudged with dirt, as were her hands and cheeks.

6441 took a shy step forward. "I'm 6441."

Delphi appeared between the two leaders.

"We're here to stop you," said 6441.

3227 shrugged. "We knew Delphi wouldn't do what we asked for, because I asked several times, in desperation, and it turned me down."

"It turned us *all* down," shouted a morph behind her, and her comrades began talking in agreement.

6441 gestured in confusion. "Turned you down? What do you want?"

Delphi disappeared.

Usually, 6441 could see its hologram around somewhere, but this time, it's as if it decided to take off and run. "That's funny." She put her hands on her hips in defiance. "It's been totally off its program since we've changed, and we've been paying the price for it."

"Us, too," 3227 shouted in an angry tone. "All we want are names."

"And not to be their slaves," several morphs said in unison.

3119, the Asian morph, stepped next to 3227. "We only want life, our own life, and not to be slaves and serve the humans."

"Names," 6441 repeated, "and a life. I get it." Seeing their numbers illuminate in small holograms over their heads, she entertained an idea, more on the spectrum of defeat. They didn't stand a chance at getting *any* of their needs and wants met. Those are human drives and desires – not for morphs. "We are our *own* species," she began, "so how do you think the humans will react to you waking them up early, possibly killing them and the mission, just so *you* can have names and argue about destiny?" She spoke in a mocking tone. "Well, even if we empathize – " Morphs behind her spoke in agreement. "Even *if* we understand where you're coming from, we still can't let you bust in there and possibly kill them all so that *you* can ask for what you want. Can't it wait?"

She showed them the trajectory on her wrist processor: 6 months to reach Wolf. Something didn't look right though. She tapped her forearm. The rectangular processor was dark.

Before she could tell everyone, 3227 threw down a self-made hammer that clanked and echoed through the slick corridor. "Those things *are* important!" She moved angrily toward 6441, who stepped in defiance toward her. "Don't you feel it?" she added. "Aren't you *sick* and *tired* of being numbers?" Her neck bulged with veins. "We sure the hell are!"

Through a thought-provoking pause, 6441 realized 3227 had a point. The last few times Delphi had addressed her by her number ID, she felt sick, angry, and diminished. She felt the reality of being manufactured: nanotechnology combined with hardware infused with downloads for consciousness. She folded her arms and rubbed her skin. Is this *all* we are? Downloads? Artificial organs? Liquid life pulp in need of re-energizing? Created for, and to obey, humans? *Stick everything into a mold and voila, or hocus pocus, as the humans say, and out pops a morph?*

"At least let's look inside this place," gestured 3327, and her twenty-seven comrades cleared a pathway to the partially opened doors.

Inside, 6441 could see a bright light. "I don't know."

"I give you my word." 3327 extended her hand for a shake. "We won't *touch* a pod. Let's just take a peek."

Morphs on both sides debated but then agreed with 3327.

6441 stepped towards 3327, meeting her and shaking her hand. "Okay, but I'm only agreeing to this because we'll *all* enter labs like this to greet our special human once he or she wakes up anyway."

Heaving and pulling the sliding doors, several

morphs forced them open. Wafts of air rushed passed them. "Wow, that was some wind," marveled 3227, steadying herself.

"Felt like it could have carried us away," 6441 added, brushing back her hair. "I feel better though, in my chest at least, but not my whole body, ouch!" She rubbed her neck, once soft in the fake muscles and ligaments, now stiff and aching. 3227 and 6441 walked into the domed room.

***

"Body pods as far as the eye can see," 3227 said, walking up slowly to the first pod.

6441 examined the interior. "It looks encrusted with ice in there. Can't see a thing."

"Ice would freeze them, *not* keep them in stasis," said 7110, "so that opaque white substance must be some other compound." His green eyes were fixed on the body pod, from bottom to top. Suddenly, he nearly fell backward. "Ah!"

"What?" 6441 ran to him.

"I can't – no way – "

What she saw made her eyes blur; she nearly collapsed into his arms.

At the top of the translucent body pod, drawn in bold print and filled in with color, she saw *her* face. "This can't be! No way!"

3227 backed into a body pod in the next row. Several morphs grabbed her arm as if terrified to take the slow steps forward. She couldn't look, but 7110 did. "This pod has *your* face on it," 7110 said.

"How?" 3227 wiped the lid, trying to look inside.

All the morphs began walking through the maze, each morph gradually calling out his or her body pod.

"Delphi!" 3227 cried.

6441 tried shaking the pod, under which two bases began glowing. "Ah!" she screamed.

Every morph stopped.

"Is it coming to life?" She backed away.

"The person inside must be waking up!" 3227 knelt. "Why did I come down here?" She looked about to cry. "What the hell made me think to consider coming here?"

Fear and curiosity wound through 6441, making her approach the glowing pod that suddenly hissed. "It's opening!" She stood at the cracking lid and put her hands in front of it in a warning gesture. When the pod opened completely, the lid creaked, and Delphi appeared on the other side.

Several morphs craned their necks, peering inside.

"Empty!" 6441 touched a long white cushion with a pillow for a headrest. She ran her fingers along the white cushion, speckled with a glittering substance. "Empty!" She smelled her warming fingers and saw lines form on her palms. "Ah!" she screamed.

3227 touched the nameplate. "It has your face on it. I thought your assigned human might be inside. I don't get this." She turned around in a perplexed motion, observing every little nuance. High above, computer hubs were lightly droning – processors streaming information to each base and body pod.

"We all saw them," said 7110, his black face brimming with shock. "We *saw* humans enter the body pods, then the lids closed – "

"And scientists put them in stasis," interrupted another morph who had located her nameplate and was touching the lid. "*People* should be inside these things."

"And there are supposed to be three other quads just like this!" 3227 lunged at Delphi. "Okay, enough. What's going on? Where are the humans?"

6441 gestured in frustration. "I don't think they're here, or we're supposed to clone them or something with these things," she said, hitting her pod.

In deep thought, 3227 shrugged and stroked her hair. "Well, if my memory serves me correctly, we only saw monitors that *showed* humans entering the pods. No morph had arrived in Zenith City at that time to witness a human stepping inside one of these things and falling asleep."

The other morphs agreed.

"So where are the humans who are destined to colonize Wolf in six weeks?" asked 6290.

Delphi illuminated a trajectory displaying the Wolf solar system. "We are currently in the Wolf solar system; the Zenith will arrive in orbit tomorrow."

6290 leered at the hologram. "You mean, like, right now? We're at Wolf *now*?!"

The morphs began shouting words of astonishment.

One morph threw an explosive through Delphi that popped on the floor. "You've been lying to us? Why?"

Delphi appeared in another form rows away from the man who tossed the putty bomb. "We will arrive at Wolf at seven thirty-five p.m. Earth time. I will convert to Wolf time as I assess Wolf's rotation and revolution."

Delphi walked over to 6441 and 3227. She motioned for 3227 to stand alongside the pod displaying her picture. "Your name is Donna Smith, and you are one of *Zenith*'s twenty navigators."

She laughed. "No, no way, uh-uh." She backed up. "I don't even know the shape of this city-ship."

"I can prove you are Commander Donna Smith, and you, 6441 are Commander Joyce Mary." Delphi turned and nodded at 6441.

"Me?! I have no knowledge of this spaceship either!" When she accidentally stuck her hand inside her body pod, her arm warmed. An awareness, as a flow of consciousness, flooded her thoughts into images and experiences from a past she believed she had never experienced. She stopped, enjoying the infusion. Peace settled inside her in a brief recollection of a distant identity and Earth. She also realized that she – and all the morphs – would need to learn so much more knowledge that only the body pods could provide for them. She didn't quite know what she needed or how much, but a force was prodding her to do something important, because putting her arm inside the pod had started a transformation, and she didn't have much time to complete it.

"Get her hand out of there!" 3227 cried.

"No," 6441 said in surrender. "I know what I have to do. Go inside." Quickly she jumped inside her body pod and lay down. She closed her eyes and could hear her comrades yelling and shouting, "Come out!" "Get out of there!" "Delphi, open the lid!"

In seconds, she slept, and for the first time, she dreamed. Dream, or reality?

***

Ten minutes later, the lid opened. Morphs reeled in fear, but 3227 and 7110 grabbed Joyce's hands, helping her climb outside and step on the sleek ground. She laughed. They turned away from her, but she kept giggling, and in full recognition of her surroundings. "They did it!"

"Did what?" 3227 asked, her body language exuding fear.

Joyce began touching them. "We're really here!" She patted shoulders and hugged a few bodies. "It worked!"

"*Who* did what, who's *here*, and *what* in crying out loud works?" asked 6290.

"Vicki!"

6290 looked around. "Who's Vicki?"

"*You're* Vicki Roberti, that's who. 6290 *is* Vickie Roberti, and 3227, *you're* Donna Smith. Don't you remember?" Joyce hugged Vickie. Vickie recoiled and screamed, but Joyce grabbed her arm and then pulled in

3227. "We're best friends, the three of us."

"Huh? What's a best friend?" asked 6290.

Joyce could only see her as Vickie now, and she huffed in frustration. "Look, we three graduated from the Space Institute together, became close while training on Mars, and then volunteered as experiments to colonize Wolf. We're experiments, *successful* experiments, *yippee*!"

Vickie and Donna cried, trying to scramble away from her.

Joyce finally quit trying to talk the truth to them and opened their body pods while the other morphs appeared jittery and yellow-faced sick. "Look, everyone, *I'm* here. Before, you trusted me to lead you. Now, trust me in all this. Look, I came out of the process just fine, and I'm alive. I'm human! I'm flesh, bones, and blood now!" She touched her chest and gestured for them to poke her. "It's our genetic material in all the pods." She explained that the pods healed her need for re-energizing, and the genetic material enters a body as gasses and seven simultaneous compound infusions. "Our organs were synthesized on Earth, and all these years, our internal technology has been repairing us." She showed them her bare calf. "No more marks or attachments to energy booths anymore!" She took Vickie and Donna's hands and held them tightly. "*We are the humans*!"

Their faces had void expressions.

"Trust me. Just do it." She gestured for 3227 to get inside her pod. "*You're* the one who desperately wanted a name. Here's your chance. Besides, you'll gradually do it anyway." She laughed. "I know you."

"Yeah?" 3227 – Donna – glanced askance at her pod. Her short brown hair curled athletically around her sculptured jawline. Once the transformation would be complete, Joyce knew Donna would be so happy the experiment succeeded.

"Well, get inside, and you'll soon realize that you *already* have a name *and* a destiny." Joyce felt like dancing, singing, touring the ship, and traveling to the sealed dome area where several of them could examine Wolf from a distance before colonization. "We *all* have names and a special destiny!" She felt exhausted in the explanations. "Trust me!"

Joyce had another realization. "In pre-programmed spurts, every morph will become aware as we have. They'll be compelled to transform. Zenith will set a schedule for transformations. After all, Delphi won't let the ship run out of energy," she laughed. "But now, since we know what's happening, we can lead them to their specific pods and help them understand what's happening to them without their thinking that the world is crashing down on them."

Donna wiped a tear out of her eye. "Well, OK, Joyce, if that's really your name."

"It is. Trust me."

"I'll do it." She stepped inside her pod. The lid shut, the pedestal glowed with processing power, and within minutes, 3227 emerged as Donna Smith. She couldn't stop laughing and hugging Vickie, who stepped out of her transformation pod with wild-eyed awareness and enthusiasm. For minutes, Vickie entertained everyone with her usual smart-aleck puns and musical rhythm.

7110, the African American morph, transformed into Colonel Dudley Mahoney, who immediately tried to convince his wife, 8888, that she was Rosemary, his wife. “We really are married! Step inside your pod and let it do the work to convince you,” he kept prodding her. He had to force the lit shut to stop her arguing. They all thought that would be the end to their fun play, but no, for three more hours, as the rest of their comrades transformed, like butterflies emerging from cocoons, Dudley and Rosemary couldn’t keep their hands off each other.

“Love,” Joyce nudged Donna, “when’s that going to happen to me?”

“Where’s Richard?” she wondered.

Delphi finally produced a manifest. “Richard Smith is Morph 10035, in Zone 4, and scheduled for transformation tomorrow at six-twenty a.m.”

“Yeah – I get a little freedom!” she celebrated.

“Freedom? Haven’t you had that already for decades?” Joyce winced and laughed.

***

Vickie had to wait until tomorrow for her husband to transform. Zenith City had everything pre-programmed so that the ship could imbibe dark energy and dark matter, enough for everyone to change before establishing orbit around Wolf.

Delphi gradually faded away. As the new humanity emerged with all their intelligence, skills, and aptitudes, Delphi grew less aware, less dominating, and less powerful. The mighty AI had achieved its purpose: to bring the newly evolved humans to their new world with Zenith City in one piece. Its final destiny was to be a holographic historian and teacher to future generations on, and orbiting, Wolf.

Later that day, Joyce returned to her quarters. As one of *Zenith*'s twenty commanders, she would soon relocate to the center dome arena, where she, Vickie, Donna, and the others would determine the amount of seeding needed for Wolf, then land Zenith City on the virgin world, set *Zenith* to activate all the manufacturing equipment, and help geneticists create and clone animals for colonization.

At their first meal, crew members continued to evolve and take their rightful places on *Zenith*. Commander Joyce Mary called on her favorite music, and activated the kitchen for full meal preparation and distribution.

"I can't believe you're over me as commander," Vickie said, tasting food for the first time. "Spinach with eggs, exquisite!"

Joyce chuckled and ate a piece of her quiche. "It's called an omelet, pal."

"You know I don't eat meat, right? I like growing my own veggies," Donna said in a testy voice. "I don't trust this crap that contraption's serving us." She pulled out a cooked spinach leaf with butter and pressed it on her tongue. Closing her eyes, she pursed her lips, the taste a tad repulsive. "And to think we've been living off energy for decades, not eating a damn thing." She looked angry. "The torture. It's been torture!"

Vickie laughed. "It's all over now, pal." She patted her shoulder.

"It's only because we know who we are now," Joyce said. "I'm just glad the great experiment worked, or we'd still be anthromorphs, space-folding for eternity." She closed her eyes, giving the God of the universe a grateful *thank you*. Because, after all, the universe is a mighty big place, and they could have been lost forever in it, had not every piece of the experiment fit into place.

# # #

J.P. Osterman has a B.A. in English from the University of San Diego and an M.A. in Education from Azusa Pacific University. She also studied graduate psychology and counseling at Texas State University, and uses that knowledge to infuse her characters with depth, conflict, and personality. In the 1990s, Ray Bradbury inspired her to write her brand of science fiction: "fiercely fantastic alien worlds." JP won the prestigious Rupert Hughes Award at the Maui Writers Conference for her sci-fi novel, *The Matter Stream,* which she transformed into her *Neltan* series. She has written fourteen novels. You can discover more of her work online at www.jposterman.com.

An Apollo 17 onboard photo of an astronaut beside the Lunar Roving Vehicle (LRV) on the Moon's surface on Dec. 7, 1972. The LRV was a lightweight electric car with battery power sufficient for about 55 miles. It weighed 462 pounds (77 pounds on the Moon) and could carry two suited astronauts, their gear, cameras, and several hundred pounds of bagged samples. (NASA/MSFC)

Where there is no vision, the people perish.

Proverbs 29:18

I Touch the Future, I Teach!

Cynthia Pearl

"I Touch the Future, I Teach"

Christa McAuliffe, 1986

***

On January 28, 1986, at 11:15 a.m., an announcement came over the PA system at Killarney Elementary School in Orlando, Florida.

"Challenger is cleared for takeoff. Students may go outside to watch the launch."

"Okay kids," I said. "Don't forget your sweaters and jackets. It's cold out there this morning."

The children jostled each other and formed a ragged line. I opened the classroom door, and they filed out behind me onto the grassy area between the rows of classrooms. Children were clustered in groups with their teachers all around the school campus. Excited and chattering, we stood shivering in the cold morning air, awaiting takeoff.

Christa McAuliffe, a 37-year-old teacher from Concord, New Hampshire, was going to be the first everyday person in space. As I stood outside my classroom with my students gathered around, waiting for the launch, I felt pride in seeing my profession recognized in this way. The space program had selected Christa because they wanted to present someone ordinary, someone relatable. What better choice than a teacher? She wasn't glamorous. She wasn't a movie star. She was a social studies teacher. And she had looked like a teacher as she jauntily marched out to the launch pad with her fellow astronauts, waving and grinning at the cameras. Teachers and school children everywhere were ready to rocket into space with Christa.

And then the countdown. There would be no delay this time. It was a go! In those moments before takeoff, how many other teachers across the country were thinking that it could be them sitting in that compartment, thrilled and terrified, strapped on top of more than 500,000 gallons of liquid oxygen and hydrogen, waiting to be rocketed through the atmosphere? How did Christa feel when she realized it was really going to happen? I imagined her holding her breath, counting down the seconds. A fellow teacher stood by the door of her classroom, monitoring the TV. She signaled the countdown, and we all joined in. 10, 9, 8, 7, 6, 5, 4, 3, 2, 1!!!! Blast off! A few moments and then suddenly there it was, the space shuttle Challenger rising in a clear blue sky, a plume of smoke mushrooming in its wake, pushing it higher and higher.

Christa was taking all of us into space. Look kids! You can do this! We are going on an adventure together, and I can't wait to share it with you. For 73 seconds we were off, everything was a go. For 73 seconds, school children all over Central Florida stood on the grass and cheered as the space shuttle rose, arching slightly to the right. My students shielded their eyes and pointed. Then there was a flash, and the vapor trail transformed into a huge swelling fist with two crooked fingers reaching out. Smoky tendrils like the aftermath of exploded fireworks streamed down. We stood there stunned, not quite sure of what we were seeing as it slowly dawned on us. Something had gone terribly wrong.

Just 73 seconds after takeoff, we had watched our hopes explode into smoke and debris, not yet able to grasp that Christa and her fellow astronauts were gone forever. As the smoke hung in the frigid sky, I hurried my students back into the classroom, anxious to know what had happened. I flipped the TV channels, only to see replays of what we had seen with our own eyes. The children were confused and subdued. They looked to me for reassurance. I don't know how we got through the rest of the day. We were all numb, not from the cold, but from the realization that a day that began in exhilaration had ended abruptly in horror, and then sadness. What goes through a child's mind when a rocket explodes in front of them? Does it make them a little more afraid? Are they a little less likely to take a chance? Do they decide they don't want to be like Christa McAuliffe? Was this the first time they realized real life is not a movie, endings are not always happy?

Christa didn't get to complete her mission. In the days that followed, information trickled out, and we struggled to put the pieces together. According to the reports, there was no "explosion," no "bang." The fuel tank just tore apart, and the spilling liquid oxygen and hydrogen burst into flames. The space shuttle was destroyed, but the crew compartment continued in flight, reaching an altitude of 65,000 feet before losing momentum. Two minutes and 45 seconds after breaking up, Challenger's cabin hit the surface of the water with a force of 200 Gs. Stunning reports indicated the likelihood the crew members were alive until that moment. We were horrified by morbid speculation regarding whether or not they were conscious. What had Christa and the rest of the crew experienced in those last terrifying minutes and seconds as they plummeted through the air before dying on impact with the sea and disappearing with the bits and pieces of fuselage and cabin and dreams that sank to the ocean bottom off Cape Canaveral Beach.

Christa didn't get to talk to us from space. She didn't get to ride in parades celebrating her flight into space. She didn't get to do any of that. Christa was a teacher, and she did what teachers do best. She taught us a lesson that day. Christa believed direct experience was the most effective teaching tool and described her Challenger voyage as the "Ultimate Field Trip." In the days leading up to the launch, she brought the world into the classroom and introduced her students to possibilities. Her final lesson left us with questions, but we learned. A hero can be an ordinary teacher who dares to do something great. Christa was not "just a teacher." She was the very best kind of teacher. She was ordinary and extraordinary. Christa modeled for us, leading by example. She provided an indisputable denunciation of the adage, "Those who can do, and those who can't, teach." She showed us a teacher can capture the attention and hearts of a nation. As we watched her march out there on the tarmac and wave at the cameras, a small woman dwarfed by the astronauts surrounding her, teachers across the country were inspired.

I can still see the smoky trail of the shuttle debris rising from the fist of the explosion and reaching into the sky, two fingers pointing. An everyday person had pursued a dream with determination. The fact that the dream was not realized did not make the pursuit any less heroic. It is not the doing, but the daring to do. Heroes can be afraid, but they go on, they move forward. They climb up into that capsule, and they go! Heroes touch the future, they teach!

# # #

Cynthia Pearl, a lifelong educator, works at the University of Central Florida in the College of Community Innovation and Education. She was a teacher for Orange County Public Schools for 18 years before moving to higher education teacher preparation and research in 2001. She is Co-Principal Investigator for three personnel preparation grants through the U.S. Department of Education. Cynthia has two grown sons and two grandchildren. She resides in Windermere, Florida with her husband, Ron.

Blue Origin's New Shepard booster lands during Mission NS-11 on May 2, 2019. Named after Mercury astronaut Alan Shepard, the first American to go to space, New Shepard is a reusable suborbital rocket system designed to take astronauts and research payloads past the Kármán line – the internationally recognized boundary of space. (Blue Origin)

The man with a new idea is a crank until the idea succeeds.

Mark Twain

In the Shadow of the Mayans

Ima Pastula

Space is an anonymous vastness. An aura of mystery remains since the first Mayan astrologers noticed its influence on the inhabitants of Earth. How can invisible forces, like the rays of light, heat, and the rainbow, an anomalous star, or the gases of the Aurora Borealis, affect humans?

What were they thinking, those ancients, when viewing such an array of color and substance up there? Strange thing that; so convinced of a *heavenly persuasion* were they, a whole scheme of things emerged as the calendar onto the quaint predictions called horoscopes.

Not to assume any empirical truth will come from this, one of many casual observations of such medieval alchemy – as are the assumptions of the Mayan astrologers. After all, those ancients, in their primitive way, sensed the power of the planets they noticed as stars that surrounded us on Earth. Their answer was in the powerful calendars they invented, a form of personal analysis with horoscopes, titling the system to an even more obliqueness: the Zodiac.

As an avid and curious researcher will warn you, "Do not dismiss the Mayan's ways and means of adjusting one's passionate relations to any number of so-called universal influences." You can run a tally on how many times your horoscope is right. The universe, the most potent life-force – and death-force – remains a persistent challenge.

Perhaps we are observing an abyss? *"A deep, seemingly bottomless chasm."* Are we looking down into the universe, or is it, indeed, above us? A dilemma!

***

Enter the astronauts! The mighty warriors of infinite space. Do they read their horoscope before take-off?

Why do the brave young men and women venture into the vastness of space? To lurk among the giant planets? Research has revealed the influence of a universal collection of foreign myths and legends brought forth to form the word "astronauts." Of interest to anyone curious as to the "why" of all things unknown. Chapters in the growing *journal of reveal* lead to an thought-provoking accumulation of foreign names for things that, taken together, are the origin of the word *astronaut.*

We are the grounded, awaiting the end of the astronauts' brave journey to determine who and what they may have met during their most unusual flight into space. Pondering of what real use is the accumulation of data, from such a universal place as space, to our civilization.

It is a tense wait for the safe return of the astronauts from an adventure we can only imagine. One particular worry seems to be, what astronauts? Are they returning or invading? Indeed. With thoughts of ancient alien astronauts, so far but a hypothesis, or NASA astronauts, will we know the difference? Or will we become part of a universal fraud?

*** 6:00 AM ***

Central Florida is privileged to be able to watch the rockets roar to the unknown, to see the brilliance of the jet fuel ignite, and the billows of smoke rise from the launchpad – all from the condo's back porch. Dwellers close to Cape Canaveral have the best view – they can even smell the fumes! But for the fragility of us without a spacesuit, we would ride along, into the profound, mysterious place of space.

Have you ever pondered the reality that we walk upon this Earth without a safety belt? Dwelling in space? It's such a wonder we don't float off this crusty Earth, drawn into the stronger gravity of other planets that swiftly race by ours. There is no roof, no handholds, as you meander under the moon and the stars.

As romantic as all this is, how can we do it? One must be wary of the temptation to solve the mysteries of space and ponder the anomalies as I do. Best to accept the Earth for what it is, and admire the brave astronauts (the NASA ones) for the nerve to leave it so swiftly on a most suspenseful journey.

***

What would be a Mayan's reaction to a spaceship? Surely, if ancient aliens roam around, wouldn't ancient times be the best time to start? Perhaps, seen as a movable object, such as a stray star, by the faith of an astrologer?

We are all susceptible to our subjectivity at the moment, and that is what remains the real secret: the Mayans' true profile. When asked to write one's opinioned observations, to headline the subject of space, one can but refer to the imagination. As Albert Einstein said, "Logic will get you from A to B. Imagination will take you everywhere."

For me, I plead the *imagination amendment*: forging the imagery of a non-entity, the myth of a substance we breathe, we fly planes in, and see as a most beautiful scene as it surrounds the deep blue of the infinite universe. One can point, stare, even smell the outer-spaceness of air. However, touching it, not so much as feeling it all about us, is up-close and personal. It is the one substance that sustains life invisibly; it tenders the lungs enough fuel to churn one's mortal motor.

*** 12:00 PM ***

The thermos filled, the rich aroma of *ebony sin* coffee contained, it's time to build a few peanut butter and jelly sandwiches for the wicker basket, and fit in a bag of gummy bears and other treats for the night watch. Read of a strange *galaxy show and tell* this night; will use the old quilt on the dry grass, just so, then lay flat on my back and become a Mayan-like worshiper of the heavenly bodies. Clear weather, the stars should twinkle, the view utter perfection, and, of course, the mystery of the colossal display will be solved. I am a dreamer!

Thinking, perhaps, the rare bursts that come once in a million years are what captured the imagination of our ancient stargazers. Speculation of course; what they saw of the displays must have convinced them the universe must have control, with its powerful disturbances; that it must have the power to control one's behavior, even the very thoughts we have of fear of the unknown. It is difficult to think like an ancient Mayan; perhaps we should locate a few descendants, query them as to their use of a horoscope, ask if they pay heed to the predictions for them because of their birthdate, as their ancestors must have done. Could it be passed on genetically? There, pleading the *imagination amendment* again.

Time is short; the old Grandfather clock is banging away, the first few digits leaving nine more. Midnight, always known as the bewitching hour, but will this night be an array of something of universal proportions? I wonder what the astronauts see if this phenomenon happens while they are cruising above the ionosphere? Sort of *you have to be there* to really know?

Considering the distance between the old quilt and the reality of space, what are the odds that one can see all the details. How it space configured, beyond the color and fireworks temperament of it all? What do we understand about space as the stage for *nature's show and tell*?

***

Just like one must ask, "*What is a dream*?" Here we have a similar dilemma: a transparent, invisible domain of a vaporous unknown, however mysterious, is one of the most vital substances needed to survive as an Earthling. To be *space-wise*, I checked the Internet to see how fast the Earth is traveling through space.

Consider the following:

- Earth is rotating on its axis once every 24 hours,

at about 1,000 mph at the equator.

- Earth's orbital speed around the sun is about 67,000 mph.
- The sun, as center of our solar system, orbits the Milky Way galaxy once every 225 million years.

Why don't we feel these incredible speeds while living on Earth? Why don't we get the bends or something? The Internet assures us we don't feel anything because we're traveling at the same speed! And thanks to gravity, we're not flung into space.

Hmmm, perhaps I should remain indoors tonight. With this added phenomena, could things be either speeded up, slowed down, or, heaven forbid, stopped?

*** 6:00 AM ***

The morning after the dare of partaking in a space anomaly. It's time for ceremonial coffee, black, evil, sin type.

I laid on that old quilt, conscious of the speed by which the Earth was hurtling us along, and the dangers, perhaps, of the new display of a universal tantrum. I witnessed such a display of universal power, as strong and mysterious as the legendary aurora borealis.

What a difference knowledge makes. Should I have left well-enough alone and just enjoyed whatever went on without my permission? Yes – knowing where I was and how fast I was hurtling through space put an edge on the midnight watch. But it was worth the worry.

Knowing all there is to know, I will pace in grace and

be grateful gravity kept its promise, and I was not swept off and beyond! Not being an astronaut, ancient or of NASA origin, I will be content with my horoscopes and leave all else to the professionals that be.

***

Space has proven to be an all-time favorite mystery. The Mayans claimed to have solved the meaning of it with calendars and the Zodiac of horoscopes. However, their knowledge has lost something in translation. Back to the drawing board. It seems each generation approaches the subject with astronauts, rockets, robots, and the like, and is still pondering: "*What is space*?"

# # #

Ima Pastula is a published author and a master artist. She was an adjunct faculty at Anne Arundle Community College in Arnold, MD. After her husband retired from military service, they relocated to Melbourne, FL. She is a past president (and current secretary) of the National League of Pen Women, and a member of the Space Coast Writers' Guild. She is available for lectures with writer's groups and book clubs.

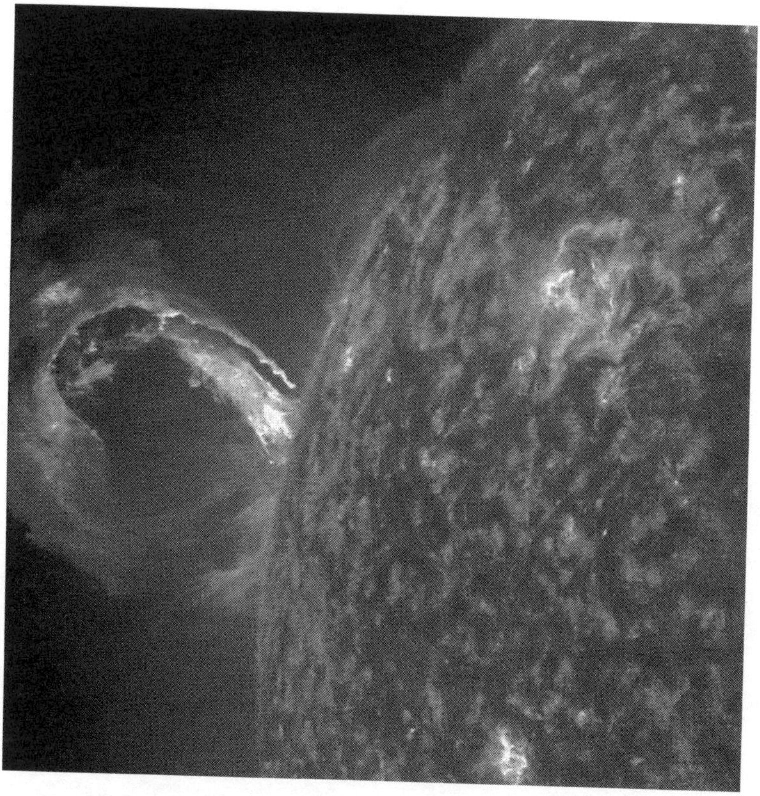

A coronal mass ejection (CME) erupted from around the edge of the sun on May 1, 2013, in a gigantic rolling wave. CMEs can shoot over a billion tons of particles into space at over a million miles per hour. Image taken by NASA's Solar Dynamics Observatory (SDO). (NASA/GSFC)

A sense of the unknown has always lured mankind and the greatest of the unknowns of today is outer space. The terrors, the joys, and the sense of accomplishment are epitomized in the space program.

William Shatner, Capt. Kirk of "Star Trek"

Voyager: The *Real* Interstellar

Scott Tilley

Voyager 2 launched on August 20, 1977, from Cape Canaveral, with Voyager 1 closely following it on September 5. I was not in Florida to see the launch in person, but I'm sure some residents will remember the blastoff of the Titan IIIE rockets. It was the start of an incredibly successful mission that continues today, over 40 years later.

Both Voyager spacecraft are still active, traveling through space billion miles from home. This is such an incredible distance that's hard for us to fathom. It takes radio signals from Voyager 1, traveling at the speed of light, over 20 hours to reach Earth. The signals from Voyager 1 are so weak that it's a monumental feat of engineering that we can even discern the signal. According to NASA, "The antennas must capture Voyager information from a signal so weak that the power striking the antenna is only 10^-16 watts (1 part in 10 quadrillion). A modern-day electronic digital watch operates at a power level 20 billion times greater than this feeble level."

The two Voyager probes took different paths through our solar system, but they both have made important contributions to our knowledge of the outer planets and their moons. Some of their instruments, such as the camera, have been shut down, but others continue to operate. This will continue until about 2025, when Voyager 1's power supply will run down. The spacecraft will continue its journey, but we won't be a direct part of it anymore.

Voyager 2, aboard a Titan III-Centaur launch vehicle, lifted off on August 20, 1977. (NASA/MSFC)

Interstellar

The movie "Interstellar," which was about astronauts traveling through a wormhole to reach interstellar space, came out in 2014. But in Voyager's case, science fact beat science fiction: Voyager 1 became the first manmade object to enter interstellar space in 2013. This was after passing through the outer boundary of our solar system, leaving the heliosphere and heading into unchartered territory. NASA named the Voyager project quite accurately: "The Interstellar Mission."

Artist's concept of Voyager in space (NASA/JPL-Caltech)

Both Voyager probes contain a golden record with images and audio from Earth. Famed astronomer Carl Sagan's voice is included, as is a message from President Carter. So far, no one has replied to our greetings, but on a cosmic scale, we've barely left home.

Voyager's Technology

When the Voyager probes were launched in 1977, the technology landscape was vastly different from what it is today. I'm typing this document on an Apple MacBook Pro that weighs just a few pounds, yet it packs a powerful computing punch: eight cores running at 2.6 GHz, 16 GB of internal memory, and a 1TB solid-state internal hard drive. I routinely enjoy over 400 Mbps wireless download (receive) speeds, and over 30 Mbps upload (transmit) speeds. And this is on a computer that's about a year old – a virtual generation in computing terms.

The Voyager probes have to make do with far less computing power: less processing, less storage, and less networking. And yet they've been able to operate for over 40 years in the harsh environment of open space, exposed to extreme radiation during their flyby of Jupiter, and the fluctuating magnetic fields at the edge of the solar system. They are truly incredible feats of engineering.

There are three different types of computers on Voyager (each probe is identical). Each computer had two versions running, usually to support redundancy, but they were sometimes used together to increase computing power. As the mission aged, this redundancy was eliminated.

The first computer is the Computer Command System (CCS), which controls the overall spacecraft. It is similar to the earlier Viking system. The second is the Flight Data System (FDS). The third is the Attitude and Articulation Control System (AACS). Each system was custom-built using components designed primarily at NASA JPL. The computers were first programmed in an early version of FORTRAN and assembly language.

Programmers today rarely think about how much memory their programs need. Even smaller applications, like those on a smartphone, basically assume infinite storage and rely on the operating system to manage things for them. The Voyager probes do not have that luxury: they must make do with just 68 KB of storage. That's about 235,000 times less memory than my MacBook.

The Voyager probes do not have any hard drive either. They rely on digital 8-track tape machines – rather like cars did for music in the late 1970s. The tape is reused when new data arrives.

Voyager 1's 8-track tape recorder (NASA/JPL-Caltech)

When Voyager 1 transmits data back to Earth, it takes over 20 hours for the signal to arrive, and when it does, it's at very slow 160 bits per second. That's far, far less than the old dial-up modems we used to use. But then again, the signal is coming from over 12 billion miles away, sent from a 22-watt transmitter. Local radio stations broadcast at 50,000 watts.

It's amazing we can still communicate with the Voyagers at all.

Voyager's Grand Tour

Once every 175 years, a cosmic alignment occurs. The four outermost planets, Jupiter, Saturn, Uranus, and Neptune, become accessible to spacecraft from Earth. The massive gravitational forces of the planets are used to "slingshot" the spacecraft forward. This means there is no need for large amounts of onboard propellant or particularly powerful engines. Nature does it for us.

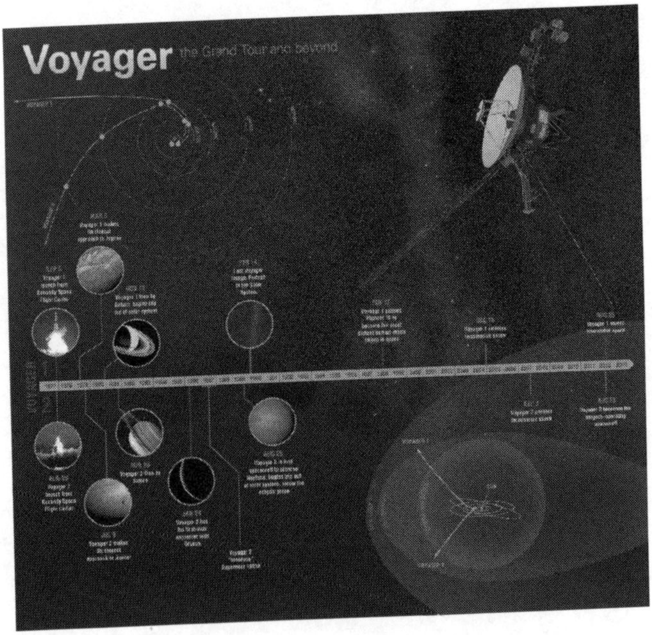

Voyager's Grand Tour (NASA)

This planetary alignment last happened in 1977, the year the Voyager probes were launched. NASA scientists started planning for this singular event in the 1960s. If anything went wrong, we would have missed our once-in-several-lifetimes chance of visiting the outer reaches of the solar system with two probes simultaneously.

Fortunately, everything went right, but there were a few glitches. According to the first "Voyager Mission Status Bulletin" published by NASA on August 9, 1977, the Attitude and Articulation Control System (AACS) and the Flight Data System (FDS) both experienced failures during testing of the VGR77-2 spacecraft – the one scheduled for launch on August 20. NASA switched to the VGR77-3 spacecraft instead, which became known as Voyager 2. This was only possible because there were actually three spacecraft assembled, but only two would fly.

Voyager 1 arrived at Jupiter on March 5, 1979, followed a few months later by Voyager 2 on July 9, 1979. After exploring the Jovian system, both probes continued on to Saturn, with Voyager 1 arriving at the ringed planet on November 12, 1980, and Voyager 2 arriving the next year, on August 25, 1981.

After Saturn, Voyager 1's trajectory took it away from the outer planets and into interstellar space. Voyager 2 continued its path towards Uranus, where it arrived on January 24, 1986. It really puts distance into perspective when you realize that it only took Voyager 2 about two years to reach Jupiter, but it took another seven years to reach Uranus. The outer planets really are "out there."

Voyager 2's final planetary encounter was with Neptune on August 25, 1989. The probe had traveled about 2.8 billion miles in 12 years, but that was really just the beginning of its journey. Like Voyager 1 before it, Voyager 2 is headed towards interstellar space. It's now over 11 billion miles from home, somewhere on the boundary of the solar system. Voyager 1 is nearly 14 billion miles away and has already crossed the heliopause into uncharted territory.

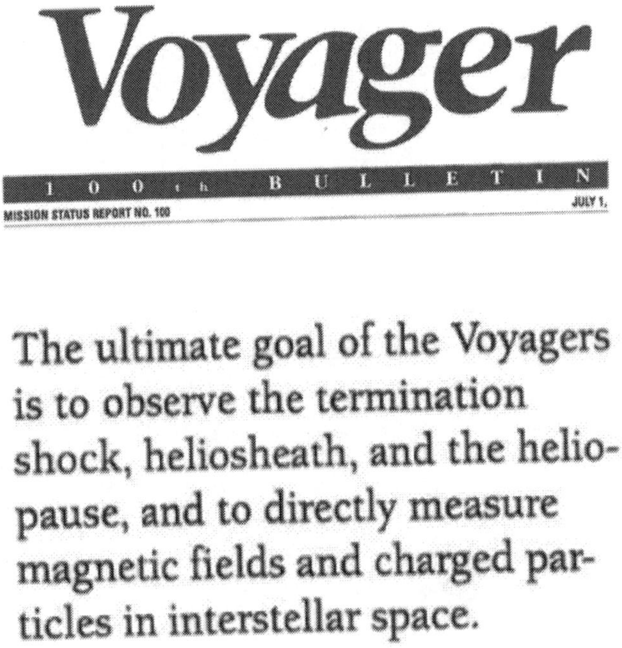
Voyager

100th BULLETIN

MISSION STATUS REPORT NO. 100 JULY 1,

The ultimate goal of the Voyagers is to observe the termination shock, heliosheath, and the heliopause, and to directly measure magnetic fields and charged particles in interstellar space.

A sample from the Voyager mission bulletin #100

In the year 40,176, Voyager 2 will finally be closer to another star than it is to our own. The probe will be

1.99 lightyears from our Sun. When we are long gone, the Voyager probes will continue their grand tour of outer space without us. Some part of humanity may someday reach the stars. Who knows, maybe the storyline from the first "Star Trek" movie will come to pass, and Voyager will return home.

# # #

Scott Tilley is an emeritus professor at the Florida Institute of Technology, president and founder of the Center for Technology & Society, president and co-founder of Big Data Florida, past president of INCOSE Space Coast, and a Space Coast Writers' Guild Fellow. His recent books include "Systems Analysis & Design" (Cengage, 2020), "Technical Justice" (CTS Press, 2019), and "Dreams" (Anthology Alliance, 2018). He wrote the "Technology Today" column for FLORIDA TODAY from 2010 to 2018. He holds a Ph.D. in computer science from the University of Victoria. Visit his author website at http://www.amazon.com/author/stilley.

This self-portrait of NASA's Curiosity rover shows the vehicle on the Vera Rubin Ridge of Mars. The photo was taken on January 23, 2018. Poking up just behind Curiosity's mast is Mount Sharp, photobombing the robot's selfie. Curiosity landed on Mars on August 6, 2012, and as of August 2019, it's still active on the Red Planet. (NASA/JPL)

To set foot on the soil of the asteroids, to lift by hand a rock from the Moon, to observe Mars from a distance of several tens of kilometers, to land on its satellite or even on its surface, what can be more fantastic? From the moment of using rocket devices a new great era will begin in astronomy: the epoch of the more intensive study of the firmament.

Konstantin Tsiolkovsky, Father of Russian Astronautics

SPACE RACE: ALIENS, PRESIDENT KENNEDY, AND CONSPIRACY

Jim Peterson

Note to readers: The U.S. government, widely recognized for its accuracy, efficiency, and expediency, has always denied the existence of aliens – the kind from outer-space, not the kind from outer-countries. We will flatly (because of the shape of this page) state for the record that we believe aliens *do* exist, and the government also believes this, but the government ("every once in a while for the people") has conducted a coverup about aliens, using top-secret blankets in top-secret hangars in top-secret places no one has ever heard of, such as Area 51, Nevada, The Pentagon, and The Bermuda Triangle.

***

As early as October 1957, the Soviet Union ("Roo-skees") launched the Sputnik 1 satellite, following a failed attempt to launch a Beatnik. A couple of months later, the U.S. ("Yang-kees") launched the Explorer 1 satellite into the *exact same* space. Ever since then, the U.S. and Russia – and later, the Chinese, French, Indians, and Elon Muskians – have openly admitted that there *is* a space race. This clearly means two things: (1) aliens are routinely racing in space, and (2) various earthling government representatives (astronauts, cosmonauts, Argonauts, and Whatnots, including people, monkeys, dogs, and guinea pigs) have traveled to space to observe, and probably participate in, the races.

Using either deductive or inductive logic, one can easily draw, or at least sketch, one's conclusions about aliens (assuming one has paper and pencil handy; do not use deductive or inductive logic near water). The reader does not need to rely on any governments' misinformation program that simultaneously maintains that a space race *does* exist, but aliens do *not*. Millions of people believe that aliens exist. In fact, more people believe in the aliens than they do in the government, mainly because more people have actually seen aliens in person, whereas they have rarely ever seen their government representatives, except possibly in a TV commercial around election time.

Even the most casual observer realizes that spendthrifts like members of Congress would not approve and fund a program to spend a gazillion dollars, work more than 60 years, train hundreds of astronauts, hit golf balls on the Moon, drive Radio Shack RC cars on Mars, station the L. Ron Hubbard Telescope in orbit, and build an International Space Station (IHOP) – with various nations like China, Spain, and Texas – just to pick up some rocks and invent Velcro. Okay, it *is* the government we're talking about, so maybe that does makes sense, but it makes even more sense that the government would do all that stuff because they are in cahoots (Latin for "secret meetings") with extraterrestrials (Latin for "more than one extraterrestrial").

Of course, the government can't just come right out and say: "Good evening, fellow citizens. We want to inform you that we are in cahoots with extraterrestrials." Such an announcement would cause uncontrollable panic, not seen since the distressing Cabbage Patch doll shortage of Christmas 1983, wherein "anything for my Susie" parents – demonstrating the *true* spirit of the season – attacked one another during festive Yuletide riots (really) at K-Marts. If the government actually said that aliens exist, good citizens would start wearing aluminum foil hats and spend too much *unproductive* time learning to speak Klingon, rather than spending so much *productive* time watching ESPN channels 1 through 42 and "So You Think You Have Talent *and* Can Survive On An Island *and* Can Dance *and* Can Be An Idol *and* Are You Smarter Than a Fifth Grader?"

Concerned for its citizens' safety and security – along with believing that most citizens are *not* smarter than fifth graders – the government obviously has no choice but to keep any alien encounters secret, and therefore to employ an official double-secret probation technique called the coverup. The method was, and still is, most widely used by Mary Kay and Revlon, but the government isn't too concerned with cosmetics, except when used on animals, most of which don't even *wear* makeup. But the government *is* concerned with presidents of the United States, particularly any who are young, smart, handsome, Catholic, and choose to live in Camelot, rather than in Washington, D.C. Of course, that president was John Jack "PT-109" Fitz Gerald Kennedy. Tragically, on November 22, 1963 (it was a truly terrible day for America and the world), President Kennedy was assassinated in Dallas. Ever since that fateful day, many Americans believe there was a conspiracy behind the president's death, and that the government conducted a coverup.

Unfortunately, many Americans did not trust the government, and people believed the government was certainly capable of conducting secret operations and then covering them up. Most Americans were familiar with the infamous 1947 Roswell incident, where everybody knows the government orchestrated a massive coverup of a crashed alien flying saucer to convince people that it was just a regular old weather balloon; at least they didn't claim that it was swamp gas! The Roswell coverup set the standard for ensuing mistrust of the government, especially when it came to matters about space and anything to do with humans traveling to or in space, or any non-human beings who were probably traveling around in space. After all, it was only sixteen years after the Roswell incident – with the subject still very fresh in everyone's mind – when President Kennedy was sadly taken from us.

It is completely understandable that most of the public did not believe President Kennedy was killed by only one man – a lone gunman. Claims of a conspiracy and a government coverup took hold immediately. The various conspiracy theories involved Vice President Lyndon "Daddy Bird" Johnson, The Mob, Jimmy Hoffa and the AFL-CIO, the CIA, the FBI, the Black Panthers, Siegfried and Roy's white tigers, the Russians and the KBG, "WKRP in Cincinnati," Fidel Castro and Cuba – to name just a few. Some Americans even believe that President Kennedy's assassin was the grassy knoll or the Texas School Book Depository building! Officially, the government concluded that the president was shot by one, and only one, man – either John Wilkes Booth or Lee Harvey Oswald. (An expert criminal profiler noted that the shooter would be a person who used "at least three names.")

A quick investigation by the astute Warren Commission eliminated the first suspect, mainly because he had been dead for almost a century. Therefore, Oswald *had* to be the man. When arrested, Oswald did not help his case when he yelled – in Spanish and Russian – nonsensical comments like: "Hey, no way I could have pulled this off solo! Just watch the 'Kennedy Conspiracy' shows on the *History Channel*!"

***

The Roswell incident, and the tragic assassination of President Kennedy, solidified for many Americans a distrust in the government and a stalwart belief that the government could, in fact, be involved in both conspiracies and coverups.

A certain percentage of the public believed – and still believes – that the Roswell incident and President Kennedy's untimely demise were actually indirectly, and possibly directly, related. The theory is that U.S. presidents are briefed about the most important ultra-secret government projects and activities, and that President Kennedy was well aware that a UFO and alien beings were discovered at the Roswell crash site, and the remains of both were housed in a likewise ultra-secret government facility. For some, that belief led to a rather short leap to concluding that UFOs, aliens, and outer space were the real reason behind President

Kennedy's affinity for, and urgency about, the U.S. space program. In short, he wanted to get Americans into space to find out what was up with aliens. Even though aliens may have been visiting Earth since at least 1947 – and most likely for centuries or millennia – it seemed entirely plausible that outer space would be a pretty good place to run into aliens. And one place that aliens might be more inclined to visit was the Moon, mainly because it was closer to outer space and there was less traffic.

Why else would President Kennedy have been so adamant about putting a man on the Moon? And why else would he want to beat the Soviets? Clearly, the president had the truly visionary perspective that the Moon was the closest place to Earth where there might be aliens, and he probably had been briefed that the Soviets thought the same thing! After all, why would the Soviets also be so interested in outer space and the Moon? Does anyone really believe that humans wanted to go to the Moon for the main reason that President Kennedy famously proclaimed in September 1962: "We choose to go to the Moon in this decade and do the other things, not because they are easy, but because they are hard"? Going to the Moon just because it is hard to do?! But of course he couldn't come right out and say, "We choose to go to the Moon because we hope to meet some aliens up there." That would have given away the big secret: the president and the government knew that aliens exist, and that they were sure of that because of the UFO and alien(s) discovered at Roswell!

Unfortunately, the president did not live to see his

bold vision completed. As we know, the president's life was abruptly taken just thirteen months after he delivered his "To the Moon" speech. For some, it seemed that there was a strange coincidence between the Roswell incident, the president's interest in space, and the loss of a great leader so soon after his declaration to land a man on the Moon. Some people felt there was no coincidence, but rather a conspiracy to prevent the president from moving forward with his bold initiative to go to the Moon in less than eight years. Many people thought a conspiracy was enacted, and then a coverup was put in place. Of course, if either the conspiracy or the coverup were true, or both, they did nothing to stop the progress of President Kennedy's principal legacy regarding the ultimate travel from this planet – the establishment of NASA's Human Space Flight Program, dramatically accelerating the space race.

***

The Soviets beat the United States by putting the first man in space in 1961. The United States beat the Soviets by landing the first man on the Moon in 1969. There have been six operational space stations – one currently. Five hundred thirty-six humans have traveled in space. There are almost four thousand satellites. Thirteen countries have space programs. Multiple countries are working towards landing on the Moon. Rovers have explored Mars and at least one spacecraft has left the solar system. Several countries are planning missions to Mars. Thousands of space tourists have paid for tickets. Almost $3 trillion has been spent on space projects. A Tesla sports car is hurtling through space.

Vroom! The space race is alive and well. Aliens are no doubt amused, watching the race.

# # #

Jim Peterson is a poet, essayist, and short story writer who is currently authoring a book about West Point's class of 1978. A former U.S. Army officer, international business executive, and college and high school English instructor, Jim resources his diverse background to creatively present readers with poignant and relatable experiences about the human condition. He lives with his wife Maria in Melbourne, on Florida's Space Coast.

Artist concept of the NASA Space Launch System (SLS) Block 1 configuration in flight. The SLS is an advanced, heavy-lift rocket that will provide an entirely new capability for science and human exploration beyond Earth's orbit. (NASA/MSFC)

We shall never forget them, nor the last time we saw them, as they prepared for their mission and waved good-bye and slipped the surly bonds of Earth to touch the face of God.

President Ronald Reagan, addressing NASA employees following the tragic loss of the Challenger 7 crew on STS-51-L

About the Editor

Scott Tilley is the president and founder of Precious Publishing, an emeritus professor at the Florida Institute of Technology, and a Space Coast Writers' Guild Fellow. He wrote the weekly "Technology Today" column for FLORIDA TODAY from 2010 to 2018. He holds a Ph.D. in computer science from the University of Victoria. He has published 25 books. Visit his author website at http://www.amazon.com/author/stilley

Anthology Alliance

Anthology Alliance

The Anthology Alliance publishes collected volumes of work edited towards a unifying theme. If you are a writing group, a non-profit organization, or a school looking to assemble and publish a collection of your members' writing, contact us! We also work with aspiring authors to edit and publish short story collections, memoirs, and essays.

Anthology Alliance is an imprint of Precious Publishing. Precious Publishing specializes in taking your writing ideas from conception to fruition. We know that your stories are precious to you, and we'll do everything we can to help see your work published.

All of our books are available online from Amazon.com in both print and Kindle formats. You are the author, we are the editor and publisher, and the world's biggest bookstore is the global distributor.

http://www.PreciousPublishing.biz/AnthologyAlliance

Made in the USA
Monee, IL
05 November 2026

67387021R10139